Skippers of the Sky

Illustration by William J. Wheeler

Front cover painting by Jim Bruce
The Curtiss HS-2L flying boat G-CAAC of Laurentide Air Service
Limited flying over the Charlevoix region of Quebec. Originally named
La Vigilance, the aircraft entered service in Canada when it was delivered
from Dartmouth, Nova Scotia, to Lac-à-la-Tortue, Quebec, on 5 to 9
June, 1919, by Canada's first bush pilot, Stuart Graham (1896–1976) and
crew, wife/navigator Madge and engineer Walter Kahre. In September
1922, G-CAAC was lost in a nonfatal crash at Foss Lake, Ontario.
A reconstruction of this aircraft can be seen in the
National Aviation Museum in Ottawa.

SKIPPERS OF THE SKY

The Early Years of Bush Flying

Selected and edited by
William J. Wheeler of the
Canadian Aviation Historical Society

FIFTH
HOUSE
PUBLISHERS

Front cover painting, Curtiss HS-2L flying boat G-CAAC, courtesy Jim Bruce, photographed by René de Carufel.
Interior illustrations by William J. Wheeler.

Cover and interior design by John Luckhurst / GDL.

The publisher gratefully acknowledges the support of
The Canada Council for the Arts and the Department of Canadian Heritage.

THE CANADA COUNCIL | LE CONSEIL DES ARTS
FOR THE ARTS | DU CANADA
SINCE 1957 | DEPUIS 1957

We acknowledge the financial support of the Government of Canada through the Book Publishing Industry Development Program for our publishing activities.

Printed in Canada.

00 01 02 03 04 / 5 4 3 2 1

CANADIAN CATALOGUING IN PUBLICATION DATA

Skippers of the sky

  ISBN 1–894004–45–0

  1. Bush flying—Canada, Northern—History. I. Wheeler, William J., 1930–
TL523.S54 2000   629.13'0971   C99-911319-4

Editor's Note: Rather than interrupt the narratives in these stories, we have chosen to retain the language of the storytellers, which reflects the time in which they were written. Metric conversions have not been added.

Published in Canada by
Fifth House Ltd.
Calgary, Alberta, Canada

Published in the U.S. by Fitzhenry & Whiteside
121 Harvard Ave.
Suite 2
Allston, Massachusetts
02134

1-800-387-9776

Contents

Acknowledgements

I would like to offer my warmest appreciation to a number of people. My list includes not only the authors who allowed me to use their work and the next of kin of those who are no longer with us but many others who helped in various ways. My thanks to Gord Ballentine; Bob Bradford, who wrote the Foreword; Doug Anderson, Jim Bruce (and the National Aviation Museum) for his striking cover painting; Ted Burton, son of the late Lucille Burton; Colin Caldwell Jr., son of the late Jack Caldwell; Tony Cashman, who has helped with terminology; Charlene Dobmeier, my patient, capable, and enthusiastic editor; Ann Sullivan, for proofreading; Bob Fowler; Bill Fox, son of the late Tommy Fox; David Godfrey, for his assistance with glossary definitions; Bob Grant and Harold Humphrey; Jean Limming, daughter of the late Tommy Fox; Larry Milberry, for fact-checking; Val Moll, daughter of the late Jack Hunter; the late John Beilby and the late Ken Molson, for photographs; Bert Phillips; Ed Rice, for his comments; Elinor Smith, wife of "Smitty" Smith; Jim Taylor; George Topple, for scanning stories; Art Wahlroth; Pat Wheeler, my wife, who suggested most of the terms in the glossary; Alan Williams, for his digital communication assistance; Alan Wingate, for providing glossary information; and Ron Wyborn, son of the late Jeff Wyborn.

1

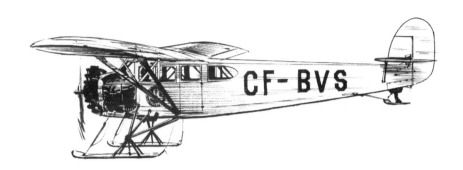

Foreword

In 1842, the celebrated English poet Alfred Lord Tennyson momentarily turned his thoughts to the vast ocean of air that surrounds our planet when he wrote:

> For I dipt into the future, far as human I could see,
> Saw the vision of the world, and all the wonder that would be;
> Saw the heavens filled with commerce, argosies of magic sails,
> Pilots of the purple twilight, dropping down with costly bails;
> Heard the heavens filled with shouting, and there rained a
>   ghastly dew
> From the nations' airy navies grappling in the central blue;

Curiously, these prophetic lines embraced the principal elements that created the subject of this book—bush flying. Indeed a number of those men of the "airy navies" who survived the "ghastly dew" of World War I returned to Canada to become "pilots of the purple twilight" in the Canadian wilderness. Joined by their civilian counterparts, they became world leaders in the art of bush flying in northern climates. It is not surprising then to learn that the first commercial bush flight in the world took place in Canada in 1919 and that the pilot, Stuart Graham, flew with the Royal Naval Air Service in World War I. The saga of the legendary bush fliers was just beginning to unfold.

The bush pilots and their companion adventurers, the air engineers, would weave a tale of resourcefulness, determination, and courage that has endured to this day. But, as famed bush pilot "Punch" Dickins once said, "We didn't think that it was unusual at the time." Our contemporary bush fliers, whether in fixed-wing aircraft or helicopters, have been inspired by those who showed the way. Much of our airline industry had its roots in the achievements of the bush fliers—the remarkable story of bush pilot Max Ward and his world-class airline Wardair is an outstanding example of this transition.

Bill Wheeler, a vice-president of the Canadian Aviation Historical Society (CAHS) and editor of its *Journal,* has astutely selected 12 first-hand accounts from the *Journal* that will provide a vivid cross-section of this colourful era in our aviation history. The reader will be drawn into a world that rivals the boldest of fiction, a world made all the more fascinating because it is a reality.

Robert W. Bradford, CM

Robert Bradford (Bob to his friends), long regarded as the dean of Canadian aviation artists, was the director of the National Aviation Museum from 1967 to 1992.

Introduction
This Book Is about Bush Flying

Bush flying, broadly defined, covers most forms of northern aviation, including aerial survey operations. It encompasses activities over the timbered regions—which diminish and end at the treeline, more or less bisecting North America latitudinally—the Barren Lands beyond, and even the Arctic. A brief inspection of the map of Canada reveals that the most populous regions lie in a comparatively narrow strip running parallel to our southern border. North of this band, roads and railways grow increasingly scarce. A few far northern communities are supplied by water transport for part of the year and by tractor train over winter roads. Flying, however, has become the only practical means of year-round access. The domain of the bush pilot and his long-time partner, the air engineer, is a vast one, including at least two-thirds of Canada. Their stories, told first-hand, are the stuff of this collection. Originally published in the *Journal* of the Canadian Aviation Historical Society, they have been selected to reflect the changes in the nature of bush flying over the years as well as the diverse conditions found in different parts of this country.

In 1929, James A. Wilson, controller of Civil Aviation in Canada, prepared a lengthy and well-illustrated article entitled "Gentlemen Adventurers of the Air" for the *National Geographic,* acquainting American readers with the principal flying activity of their northern neighbours. Intentional or not, its publication marked 10 years of wilderness aerial operation in Canada. Wilson recounted the substantial accomplishments of that first generation of bush pilots and air engineers; their confidence and optimism pervade his text. He equated these "skippers of the sky" with the determined sea captains of earlier centuries who probed the most remote corners of their world and, by so doing, Wilson fostered the romantic image of the bush pilot. But his colourful analogy was an apt one. Northern Canada was then as vast and as uncharted as any sixteenth century sea. Maps of the seemingly endless northern wilderness—when they existed—were crude and often so misleading as to be dangerous.

Wilson was not the first to see the men who flew north—disappearing completely for weeks or months on end—as dashing adventurers. Canadian journalists, largely from western Canada, had been creating flying icons since the first tentative northern aerial ventures, notable among which was Imperial Oil's attempt in 1921 to establish an air route along the Mackenzie River to Norman Wells in the Northwest Territories. Their stories of daring mercy flights were widely read and often picked up by papers in eastern Canada. Deservedly, the names of the men they lionized would become well known, joining the ranks of Canada's World War I flying aces. Most, if not all, of that first generation of bush pilots and air engineers had obtained their aviation training with the Royal Flying Corps or Royal Naval Air Service, precursors of the Royal Air Force. Many were decorated veterans who had proven their flying skills over the trenches of France or on anti-submarine patrols off the coasts of England. "Doc" Oaks, "Punch" Dickins, Stuart Graham, "Wop" May, Don MacLaren, and Roy Brown are names that come readily to mind.

Canada's first bush pilot, it is generally accepted, was Stuart Graham, who collected an ex-U.S. Navy Curtiss HS-2L from Halifax in June 1919 and brought it to Grand'Mere, Quebec, where he flew it for the St. Maurice Forestry Protective Association (later Laurentide Air Service). The Curtiss, named *La Vigilance* and registered as G-CAAC (and reproduced on the cover of this book), is mentioned in Jack Caldwell's account. Happily this historic machine, Canada's first bush plane, is now represented in the National Aviation Museum in Ottawa. Its wreckage was recovered in 1969 from Foss Lake in northern Ontario where it had been abandoned after a 1922 crash. The remains are displayed beside a faithful restoration of G-CACC, built by NAM staff, containing more than 70 percent authentic HS-2L parts. It is a unique exhibit and one of the most impressive in any of the world's aviation museums.

The aircraft that bush pilots flew over the years are fascinating in themselves. Along with the HS-2L, another World War I aircraft used extensively by postwar civilian pilots for flying instruction and barnstorming was the open-cockpit Curtiss JN-4 (Can) training biplane named the Canuck. A Canadian version of the well-known U.S.-built Jenny, the Canuck was fitted for one season with skis for winter operation into the newly discovered gold fields of northern Ontario and Manitoba—providing a timesaving taxi service for prospectors and miners. The previously mentioned HS-2L, a large biplane flying boat designed as a maritime patrol bomber, was in common usage across Canada, notably by the RCAF and the Ontario Provincial Air Service. By later standards, its 74-foot span and attendant maze of struts and rigging made for a most ungainly bush plane. Its crews accepted the challenge of tying the HS-2L to a heavily wooded shore and struggling to load passengers or supplies into its open cockpits because they realized it was the best aircraft available. Soon, smaller and more efficient machines joined the H-boat. Canada, being a land of lakes and rivers, meant that these were also biplane flying boats, notably the Vickers Viking, Boeing B-1, and Canadian Vickers Vedette.

Within a decade, almost all of the HS-2Ls were gone; "boats" were largely superseded by new utility transports, notably those built by Fairchild, Fokker, and Bellanca—fitted with floats for summer operation and skis in winter. On floats, the high-wing monoplane configuration of these new aeroplanes afforded much easier cabin access from docks or from a wooded shore. Variations of all three were built in Canada. Only the RCAF persisted in the use of flying boats on civil operations (the Service's principal prewar function), using larger and more advanced types such as the Canadian Vickers Varuna and Vancouver. Since these boats operated only from water and had to be laid up in winter, the RCAF added Bellanca and Fairchild aircraft for year-round use.

The other noteworthy bush transport of the 1930s and immediate postwar period was the then-revolutionary low-wing Junkers W. 33/34, built in Germany. Unlike its fabric-covered, wood and metal contemporaries, it was constructed of metal, making it less susceptible to the ravages of Canadian weather. Smaller aircraft such as the Stinson and Waco lines of cabin machines handled the lighter air-taxi work.

The Noorduyn Norseman, which first flew in 1935, incorporated suggestions from bush operators, and was the first Canadian aircraft designed specifically for northern conditions. It features prominently

Junkers W.34

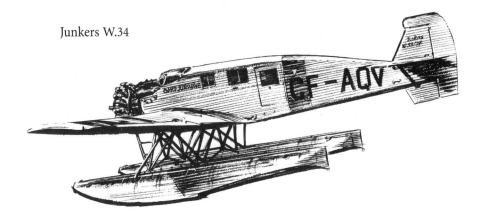

in several of the following stories. The Norseman's payload was some-what more than that of a half-ton truck. With less fuel, it could carry additional cargo. In 1947 de Havilland Canada test flew its famous Beaver and, in 1951, the bigger Otter, both to specifications provided by potential users. Unlike the Norseman, these machines are con-structed entirely of metal. In size, the Norseman is midway between the Beaver and the Otter. Almost a half-century after its introduction, the Beaver remains in wide service throughout Canada. In fact, with rebuilding and the conversion of imported ex-military machines, its numbers are increasing. In the postwar years, surplus RCAF, light, twin-engine aircraft such as the Canadian-designed and -built Avro Anson V (on skis or wheels) and the Beech 18 Expediter (on skis, wheels, or floats) proved useful in the bush, emulating the HS-2L of the 1920s. Cessna aircraft and the Canadian-designed Found largely replaced Wacos and Stinsons.

When still larger, twin-engine aircraft were needed, surplus World War II machines such as the Consolidated PBY Canso amphibian and Douglas DC-3 Dakota joined the Norseman and the Beaver. While the "Daks" and Cansos both operated as transports, the latter also flew on aerial mineral survey operations, as recalled here by "Smitty" Smith. For aerial photography such exotic types as the Lockheed P-38 Lightning and de Havilland DH 98 Mosquito—both wartime fighter-bombers—flew at high altitude taking the photographs that would translate into today's meticulously detailed maps. Bob Fowler describes this highly demanding form of flying. Commercially oper-ated four-engine Boeing B-17 Flying Fortresses and RCAF Avro

Lancasters also performed survey photography. While these wartime bombers did not fly as high or as fast as the P-38s and Mosquitoes, their crews were larger, the flying less strenuous, and the aircrafts' endurance greater.

Aircraft engines have improved in efficiency and reliability, hand in hand with the development of the aircraft they power. The massive water-cooled Liberty engine disappeared with the HS-2L; lighter and more easily maintained air-cooled radial engines became the norm. The first of these was the Wright Whirlwind, followed by the Pratt & Whitney Wasp in ever larger and otherwise improved variants. Versions of the Wasp powered the Fairchilds, Fokkers, and Bellancas, and even the Norsemans and much-later Beavers and Otters. Cessnas and other lighter machines used efficient, air-cooled, horizontally opposed Continental and Lycoming power plants. Liquid-cooled aircraft engines made a brief reappearance in the North—Rolls-Royce Merlins in the fore-mentioned Mosquitoes and Lancasters and Allisons in the P-38s. Recently, Beavers and Otters have been converted to turboprop power using the Canadian-designed Pratt & Whitney PT6.

Instrumentation in the first bush planes was minimal. Modern pilots, with induction compasses and global positioning to assist navigation, are unlikely to lose their way. Their 1920s and 1930s predecessors had only magnetic compasses to guide them, and, in the far north, this instrument was notoriously unreliable because of the relative nearness of the magnetic north pole. The deviation between

Lockheed P-38 (Lightning)

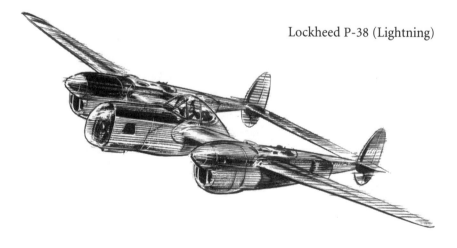

real and magnetic north could be extreme. And contemporary maps were also of marginal use. The completeness and accuracy of modern charts would have gratified any old-time bush pilot accustomed to very basic maps with few established locations and vast blank areas. When they could, early pilots might sketch useful topographical features into these open spaces and add written notations—based on their own observations and those of fellow pilots.

While today's bush pilots are in constant radio contact with their base, such was far from the case in the early years. Radios were too heavy, bulky, and limited in range to be practical aboard aircraft. Away from base, pilots and air engineers were on their own. Only if they became overdue or did not show up at a predetermined location would there be concern for their safety. Forced down, they could count on an unplanned wilderness experience, possibly a lengthy one. *Black boxes* that activate automatically to signal the location of a downed machine were far in the future. Since relatively few aircraft were available, air searches were difficult to mount and essentially hit or miss. With so much country to cover, the latter was not uncommon. In a few unfortunate instances missing aircraft were located too late to save the lives of those on board. Today, Canadian Armed Forces search-and-rescue (SAR) units, assisted by aircraft flown by volunteer private pilots, members of the Canadian Search and Rescue Association (CASARA), are activated within hours—even minutes—of a forced landing or a crash.

The introduction of the JN-4 (Can) and HS-2L, primitive as they were, to wilderness flying began a revolution in northern travel. In his article, Wilson cited instances of air transport shortening a season or more of strenuous travel by canoe or dog team into a matter of weeks by air, or a difficult two-week hike into a half-hour flight. Such savings in time and effort appealed enormously to wilderness travellers and residents of isolated communities. Determined crews more than compensated for the shortcomings of aircraft and engines. Any initial misgivings northerners may have had gave way to enthusiastic acceptance of this most modern invention. Although accustomed to otherwise primitive living conditions, they soon came to accept air travel as readily as any southerner might a bus trip.

What did bush planes carry? Old photographs show cases of obvious staples such as fresh fruit and vegetables, canned goods, cough medicine, tea (once the preferred drink of the North), and whisky waiting to be loaded into an aircraft, not to mention sacks of flour, salt, and sugar. Anything a customer might want and that fit into the cabin of an aircraft was delivered by air. While a "Cornflakes" load might fill every cubic inch of space, a shipment of concentrated ore could leave the aircraft seemingly empty.

A few cargoes, for readily apparent reasons, have become legend. A mink on its way north for a breeding experiment escaped from its cage and, before reaching the cockpit, ate its way through a suit of clothing destined for a bishop. Conscious of movement behind him, the pilot turned his head—to gaze into a tiny set of beady eyes. A load of squealing piglets left an odoriferous reminder of their presence that lingered in the aircraft long after their departure. Trucks and even small bulldozers were taken apart and cut into manageable sections, then welded back together after delivery. Mining equipment received similar treatment. Once, a sedated ox arrived via Canadian Airways' giant Junkers Ju. 52/1m, a larger relation of the W.34. All pilots and air engineers, including the contributors to this book, have a wealth of amusing cargo stories. Read on and you will learn of some unusual loads and unlikely passengers.

What qualities typified bush pilots and air engineers? Hardiness and determination, certainly. Despite the romantic perception newspapers and magazines depicted, their lives were extremely strenuous. But routine dangers and setbacks were taken in stride—almost cheerfully. And, nearly without exception, these pioneers of the air shared a ready sense of humour. Many who worked with them have commented on their unfailing ability to see the amusing side of difficult situations—an ability that undoubtedly sustained them. Humour is a thread running prominently through each of the following accounts.

Resourcefulness was another quality common to those who flew in the North. The late A. E. de M. (Jock) Jarvis, who unfortunately never did write of his bush flying experiences, was an entertaining raconteur, not afraid to laugh at himself. One of his stories dated from the late 1930s when he operated his own one-man air service out of a base east of Sault Ste. Marie (the Soo), Ontario. Initially, he flew an ancient Curtiss Robin (with its notoriously cranky Curtiss Challenger engine), which he subsequently replaced with a more modern Waco Standard cabin biplane (Jacobs-powered). In the rugged, near-mountainous country north of the Soo, Jock had set up several fishing camps that could be reached only by air. These he rented to fishermen, flying them in for as long as they wanted to stay.

He was returning from one such trip when his engine quit. Nosing down, he managed to stretch his glide just far enough to reach a tiny unnamed lake, brushing his floats through the tops of the surrounding trees. He was down safely—but no one knew where he was. His immediate well-being and that of the fishermen back at his camp depended solely on him.

Drifting the Waco nose-first into shore, Jock tied it securely to the trees. Soon he was standing on spruce poles lashed together across the front of the Waco's floats, dismantling the engine. In the crystal-clear water below, he could see the gravel bottom only a yard or so beneath him. By the time he located and corrected the problem, he had taken one cylinder completely apart. He was reassembling it when a pushrod slipped from his fingers and splashed into the water. Jock wasn't immediately concerned—the bottom was clearly visible. A 10-inch shaft of gleaming steel should be easy to see. But it wasn't. Wading and diving, he scoured the area beneath the aircraft—and found nothing. He did see a few lurking pike and wondered if one of them had swallowed the missing rod, mistaking it for a young trout. Along with the usual survival equipment, Jock carried the customary spares—fabric, dope, spark plugs, and a cylinder "pot" (piston)—but no pushrod. His situation was grim indeed.

Waco Standard

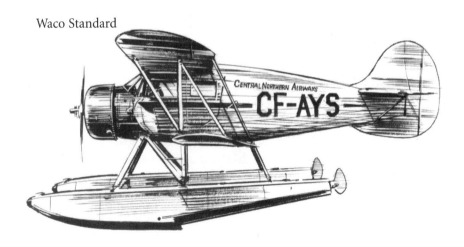

Happily, Jock was nothing if not resourceful. From his box of emergency tools he selected the largest screwdriver and, with a hacksaw, cut the shank from the handle. He now had a hardened-steel shaft of approximately the correct diameter and length for a new pushrod. But it required perfectly hemispherical ends. Now what?

Jock's ingenuity did not fail him. He loosely wired the screwdriver shank to a pair of saplings that grew close together, then bent a third sapling and attached a length of twine to it. The rest of the twine he wrapped around the centre of the shank. As he slowly released the sapling, it straightened, pulling the cord and spinning the shank within the wire that held it. Jock had devised a primitive lathe. By holding a file against one end of the shank, he was able to shape it into a perfect hemisphere. Repeating the process on the other end, he arrived at a near duplicate of the lost pushrod. With the *buckshee* rod installed, the Waco's "Shaky Jake"—as the Jacobs engine was known colloquially—ran as smoothly as ever. Jock took off safely and flew back to the Soo with no further problems.

This story has an anti-climax. Jock, in all innocence, described the incident to a friend who just happened to be the editor of the *Sault Ste. Marie Chronicle*. Two days later the article "Pike Attacks Aircraft" appeared in the newspaper, and was later picked up by the *Toronto Star*.

Alex Milne was an air engineer with the RCAF in the mid-1920s when that service was responsible for mapping northern Manitoba and Saskatchewan and for distributing treaty money to the local Native peoples. The RCAF aircraft, an all-wood Vickers Viking biplane flying boat, was powered by a water-cooled Rolls-Royce Eagle engine suspended from the upper wing, above the heads of the four-man crew. While they were photographing near the Manitoba/Northwest Territories border, a split fuel line forced them to make an unscheduled landing. The lake they set down on was a thousand miles from any form of help, and it fell to Alex to get them back into the air.

After checking his emergency supplies, Alex went ashore and cut a willow sapling—about the length and thickness of the damaged hose—and peeled it. Then he wound soft copper wire tightly along its length. This he covered with tape and sealed with several coats of dope (lacquer). Finally, he carefully extracted the slippery willow from his ingenious "hose." Fitted in place of the damaged original, its connections sealed with more tape and dope, the improvisation seemed to work. During the trip back to their base near Winnipeg, Alex and the crew kept a nervous eye on it. But it did the job.

Not all "bush planes," as explained earlier, served in the transport role. Spurred by the Cold War in its early phases, the Canadian Government decided that our vast northern reaches should be mapped—accurately. Those huge blank spaces that had been the bane of bush pilots for generations were to be eliminated. The RCAF initiated the project using converted Lancaster bombers of World War II vintage, and, because of the urgency of the operation, commercial firms such as Spartan Air Services and Kenting Aerial Surveys received contracts as well. The former flew at very high altitude with converted P-38 Lightning and Mosquito aircraft while the latter used the B-17 Flying Fortress. Thus, in the 1950s, high performance and once-lethal military aircraft could be seen parked alongside Norsemans, Ansons, Beavers, and other more "tame" aircraft on some northern strips.

We are fortunate that stories such as those in this book have been recorded. They are, for the most part, from a bygone era; the events they describe are unlikely ever to be repeated. Where isolated communities once depended upon single-engine, float- and ski-equipped machines that landed on nearby waterways, larger multi-engine aircraft, jet- or turboprop-powered, now operate on schedule from all-weather landing strips. The much-improved equipment, facilities, and electronic aides available to modern pilots have removed many of the uncertainties of earlier times—uncertainties that could translate into episodes such as those described here. The risks that were once the stuff of bush flying are no longer the norm. For instance, the weather, while it will always be a factor, is not as critical as it once was. Pilots can now fly above it and, when their destination is "socked in," either divert to another location or let down on instruments. In the not-so-distant past the only option open to a pilot caught in a snowstorm was to land on the nearest lake while he could still see to do so, make camp, and wait it out. Some small air services continue to use lighter aircraft, but these companies also have relatively sophisticated equipment and tend to restrict their operations to local flights.

Women have been associated with aviation in the North almost since the first bush plane arrived there. Jeff Wyborn mentions two women who ran support facilities. Although Canadian women have been flying since the 1930s, until the postwar years none had become bush pilots. Vi Warren (Milstead), who had delivered aircraft to squadrons in wartime England, probably came the closest when she and her husband, Arnold, operated a flying school/air service out of Sudbury after World War II. She flew the occasional wilderness trip but saw herself more as an instructor with husband Arnold handling any bush flying assignments that came their way. The wives of pilots and air engineers did manage occasionally to accompany their husbands, sharing the hardships and loneliness of life at isolated bases. Their contributions have gone largely unsung. Air engineer Alex Milne recalled returning to base to find that a used Fairchild that his firm, Dominion Explorers, had purchased was now freshly painted in Domex colours—by hand, with brushes (no masking tape or spray gun). The trim wavered only slightly. His wife, Elin, and another woman had taken on the job to surprise their husbands. To this writer's knowledge, only Connie Dickins (in a privately published booklet) and Lucille Burton ever wrote of their experiences, and we are lucky to have Mrs. Burton's recollections for inclusion in this book.

Bert Phillips remarks that bush flying was (and is) a day-to-day business. It could be as prosaic as driving a pick-up truck. The more regular the routine, the better it suited the aircraft owners. But routines seldom went unbroken. And it is the unexpected—the emergencies—that are the spice of these stories. The drop-everything mercy flight, an unruly passenger, a sudden loss of power, a damaged undercarriage, wing icing, a bird strike, frozen instruments, an engine fire, unusually turbulent air—the list of potential hazards, all memorable in their resolution, is endless. Even landing at a tricky location or getting used to a new type of aircraft could provide a challenge, adding variety and richness to the recollections of bush pilots and air engineers.

The tales that follow are not embellished. Only two or three of the writers have had any degree of writing or public speaking experience. Their stories draw substance from the fact that they were *there*. All of these accounts, following their CAHS *Journal* appearance, have earned the approval of a knowledgeable membership, many of whom have shared similar adventures. They describe bush flying as it was and, to a limited degree, may still be. The matter-of-fact understatement and the absence of artifice make their stories all the more vivid and plausible. In all of them, a love of flying—even a sense of exhilaration—is apparent. The artificial, but once popular *gee-whiz* style of writing will not be found here. Yet the pride each author took in carrying out a demanding job is obvious.

If you enjoy stories that blend history and flying, such as the accounts in this book, you may be interested in becoming a member of the Canadian Aviation Historical Society, founded in 1962. Information on the society can be obtained by writing to the CAHS at PO Box 224, Station 'A,' Willowdale, ON, M2N 5S8, or contact www.cahs.com on the web.

Among bush pilots the standard of airmanship was of the highest. Yet there was attrition. Even the most capable and experienced were not immune. Jack Caldwell, Andy Cruickshank, Fred Stephenson, Wilson Clarke, George Milham, Tom Higgins, and "Bun" Paget are only a few of the many well-known and respected Canadian pilots and air engineers who lost their lives in flying accidents. This book is respectfully dedicated to the memory of these gallant men.

William J. Wheeler

The Earliest Days of Bush Flying

COLIN S. (JACK) CALDWELL

The late Jack Caldwell was one of Canada's earliest bush pilots. His brief career, terminated by the fatal crash in June 1929 of a new Fokker Super Universal he was flying for Canadian Vickers, spanned a mere eight years. Yet his exploits have earned him lasting fame. They ranged from pioneering in aerial spotting for the Newfoundland sealing fleets in a tiny Avro Baby, single-seat biplane of World War I vintage to flying a mineral survey party deep into northern British Columbia and Yukon in a Vickers Viking five-place, open-cockpit flying boat—the first aeroplane ever to penetrate the remote Stikine and Liard regions.

Jack Caldwell was also the sole Canadian entrant in the On-to-Dayton air race of 1924, piloting a Thomas-Morse Scout, an American World War I fighter design that never reached Europe but was used instead as a trainer in the U.S. His days of adventurous flying in the farthest corners of Canada supposedly ended when he joined Canadian Vickers of Montreal as a test pilot. However, he became the first Canadian member of the famed Caterpillar Club when he was forced to take to his parachute, abandoning a new Canadian Vickers Vedette he was testing prior to its delivery to the RCAF. The only Vedette in existence, a reconstruction, can be seen in the Western Canada Aviation Museum in Winnipeg.

The Super Universal in which he died was one of a number of such machines built under a licence from the American parent firm. He flew into an electrical cable newly strung across the St. Lawrence River and of which he had not been advised. As with the HS-2L and

17

Vedette, only a single Super Universal exists in the world. And like them, it is in Canada. Clark Seaborn of Calgary recently flew this machine after spending 16 years rebuilding it.

The following account covers the beginning of Caldwell's flying career. He was piloting a U.S.-built Curtiss HS-2L flying boat—a World War I maritime patrol bomber used by American units based in Nova Scotia and pressed into service as Canada's first bush plane following the war. The HS-2L was a large biplane of wood and fabric construction with a span of 74 feet and no fewer than eight pairs of interplane struts and an attendant maze of rigging. An old joke had it that if a chicken were released within the struts and escaped, it meant something was broken or missing. A heavy, 360 hp, water-cooled V-12 engine mounted above and behind the crew—ideally located to crush them in the event of a serious crash—powered the aircraft. To modern eyes the HS-2L would seem an impossibly awkward machine for wilderness operation.

The aircraft involved in the crash Caldwell describes is the same HS-2L, registered as G-CAAC, that is represented by a unique exhibit in Ottawa's National Aviation Museum. While the NAM's HS-2L is a faithful reconstruction containing many original parts, the actual remains of the aircraft that crashed in 1922 are displayed with it.

It was three years after the war when, becoming increasingly bored with civilian life on the ground, I seized the opportunity to return to flying and joined Laurentide Air Service as a mechanic. That summer, the company was engaged in an aerial survey for the Ontario Government in the James Bay area, operating five aeroplanes. Our base was at Remi Lake, about 50 miles west of Cochrane, and we spent the entire summer over the territory, sketching. It sounds prosaic, and generally it was.

Each pilot made a daily trip of four-and-a-half hours, during which time we did the mapping and sketching. For part of the season, we operated from a base at Moose Factory, where the day's flight carried us over floating muskeg extending for 40 miles south of James Bay. It was an indescribably desolate region—nothing but oozy swamp with little patches of water here and there—not a landing place in all of its extent. As we sketched, we had to be constantly on the lookout for the

C. S. (Jack) Caldwell at Ramsay Lake (Sudbury), in 1923, during his first year as a pilot. The aircraft in the background is a Curtiss HS-2L. *G. Swartman via K. M. Molson*

next place to land, with lots of opportunity to wonder what would happen if we had to force land.

At times, there was no choice, and we had to make forced landings under difficult circumstances. Once, returning to base after surveying all day, we ran into a heavy rainstorm. After a lengthy buffeting, we were forced to set down in a small lake that had come conveniently to hand. After the storm subsided, we discovered how painfully small the lake really was. We attempted to take off, but the machine would not climb quickly enough to clear the trees on shore. We did some rapid thinking, but there was nothing to be done but turn back and hope for the best. Almost inevitably, the machine struck the water awk-wardly and began to sink. We barely managed to get clear. We had no time to recover emergency rations, matches, or anything else before the aircraft disappeared with a gurgle—leaving us to swim to shore.

Luckily, this was not far or we might never have made it. But 30 miles of practically uninhabited wilderness separated us from our base. In the darkness, the only route to follow was along the bank of a river out of the lake. All that night we forged through the blackness, practically feeling our way. It was a weary, grimy, and famished pair

Curtiss HS-2L, G-CAAC, after crashing in Foss Lake (near Kapuskasing, Ontario) in September 1922. After the mishap Jack Caldwell and the other crewmembers spent the night and the next day trudging through dense bush. Forty-seven years later the remains were raised and can now be seen in the National Aviation Museum in Ottawa. *NAC/DND RE-5938*

that stumbled into a trapper's shack next morning, still many miles from base.

On another occasion, a pilot out on the survey miscalculated the amount of gasoline he was carrying—or tried to make it go too far. He suddenly found himself 30 miles from base with a sputtering motor. The only body of water in sight was again a lake of dauntingly small dimensions. Gliding will take a machine a long way under such circumstances—but there are limits. There was nothing to do but endeavour to hit the lake. Luckily, the pilot accomplished this safely. Knowing that their aircraft would be the best means of attracting attention to their plight, the three occupants moored it in the middle of the lake and swam ashore with the emergency rations, prepared to spend the night.

Flying down from Moose Factory that same evening, we crossed directly over the lake and spied the plane. We knew the pilot would never have alighted there by choice so we came down and circled the lake several times before spotting three wildly gesticulating figures on the shore. It was impossible for us to land with their machine on the lake. And theirs was the only aircraft in our fleet possessing quick enough take-off and climbing powers to get out of such a restricted area. With dusk closing in, there was nothing we could do but continue on to base and report that one of our machines was down in a duck pond 30 miles away.

Early the following morning, I set out with a mechanic to see what assistance I could offer. Reaching the lake, I flew low and began circling. Almost immediately, the words "DROP GAS" stamped in gigantic letters into the mud of the beach caught my attention. The request made sense, so I returned to base where we secured five stout five-gallon cans and filled them with gasoline. We crated them with strong boards and loaded them on the aeroplane. Then I set out again.

Once over the lake, I flew my boat as low as was prudent parallel to the shore, and, on each circuit—at precisely the same spot—my mechanic released a can of fuel. I could see the stranded pilot standing stark naked upon the beach, eager and expectant. As each consignment hurtled down, he made a wild dash through the water after it. We had no way of determining our success, which, to me, was problematic. Even from that height and at our comparatively low speed, water is practically solid. Yet the three seemed to be signalling that all was well, and we returned to base. The missing plane came in later in the day. Two of the five containers had survived the impact, and the trio was able to gas up and get out of the lake.

FOREST FIRE PATROL

After an idle winter, I engaged for the following summer in comparatively prosaic flying work—aerial forest-fire patrol for a pulp and paper company that owned large reserves in Quebec. The season was one of the worst for fires in many years. Most of the patrol work was carried out in a thick haze of smoke, making flying difficult. We reached the end of the season with no mishap or episode sufficiently unusual to register permanently on my mind.

I recall one humorous incident, however. It pertained to another pilot involved in forest patrol work. His biggest fear, when travelling through the woods alone and on foot, was of bears. One afternoon he was forced to land and abandon his aircraft far from base, and, typically, his principal apprehension was that he might encounter one of these usually harmless animals. Accordingly, he retrieved a tin can from the disabled craft, and, as he forced his way through the dense bush hopefully toward his base, he kept up a ceaseless tattoo to scare off any bruins in the neighbourhood. Happily, his plan proved fortuitous in another way. A trapper heard the hammering and went to investigate, and, to their mutual surprise, the two came upon one another. The pilot found food and shelter long before he had anticipated and was put safely upon the trail home.

THE ROUYN GOLD FIELDS

During 1924, Laurentide Air Service established the first passenger and air express service in Canada. Strange as it may seem, this was not between any of the major population centres, but from the edge of civilization into the newly discovered gold fields in northwestern Quebec. From Haileybury on Lake Temiscaming our machine flew into the primordial wilderness of Rouyn. It must have been a flying service then unique in the world; mining authorities have since credited it with setting gold field development ahead by two years.

From our Haileybury base, we undertook to drop passengers or express in any part of the gold fields where there was a lake—a con-

Curtiss HS-2L, G-CACW, of Laurentide Air Service being manoeuvered toward shore. The man on the wing is using his weight to help turn the aircraft. The bow cockpit in just such a machine was the scene of the struggle between two drunken miners described by Jack Caldwell.
R. Vachon via K. M. Molson

dition easily met, since the area was dotted with bodies of water. At first there was no great rush of passengers, though it took a good five days of strenuous paddling (and slow consumption by mosquitoes and black flies) to reach the heart of the gold belt. We accomplished the journey in almost exactly one hour. Men who would face many kinds of death in the wilderness found excuses for not flying. But we continued to carry express of all sorts without mishap and, gradually, confidence grew, and the number of passengers increased. At the height of the season we were transporting a weekly average of 30 passengers in and out of the gold fields.

I am sure that the aeroplane has never before been requisitioned for such multifarious services or been subjected to such indignities by way of loads, as in our daily service to the gold fields. I carried every imaginable commodity: fresh meat, fruit, vegetables, eggs, dynamite by the box, horseshoes, and even dogs and cats. I once flew over 60 miles of wilderness with 15-foot lengths of heavy piping strapped to the wings. On another occasion, a 500-pound pneumatic drill was dismantled and stowed away in various parts of the plane's capacious interior and delivered safely to a Rouyn mine. Another time, we carried an entire drum of gasoline by pouring it into two containers and putting one in either cockpit. There was very little we had to turn down. The mining camps came to depend on our service to such a degree that a teamster awaiting a new team of horses became convinced that the animals would come by air.

About the only thing we did not knowingly carry was liquor. We sometimes flew it in when we had no control over the consignment. One day, two noisily happy and inebriated individuals—miners or prospectors returning from celebrating—approached me as I was warming up the machine and expressed a hiccoughed desire to be transported by air to Rouyn. As they had their fares I could not well refuse them, even had I felt so inclined. Suspecting the possibility of more liquor in their packs—in addition to what they were carrying internally—I stowed these well aft in the boat and placed the two men in the forward cockpit.

One's first take-off experience is always a bit disturbing, and this one had a decidedly sobering effect upon my two passengers. For some time they remained quiet, glued to their seats. Unfortunately, their concern rapidly wore off. Soon the two heads began periodically disappearing as they ducked down, apparently for refreshment. I was not apprehensive, however, until a whisky bottle—seeming to me the size of a balloon—whizzed past the propeller. Suddenly one of the men was on his feet, swaying unsteadily. Swinging his arms with clenched fists in the most bellicose manner, he invited the other to fight. A second later they were at it, grappling and struggling in their narrow, unstable quarters. I didn't know what to do. In fact, there was nothing to be done but to fly on as straight and steadily as possible. Fascinated, I watched as the two in front of me listed from side to side, tightly locked in each other's embrace. I expected them to topple out at any moment, as one would gain a temporary advantage and force the other dangerously back over the edge of the cockpit.

Fortunately, nature intervened. We struck a downdraft, and the machine dropped several hundred feet in that sickening manner that makes the novice think his stomach has left him. It ended with an abrupt jolt—as if the earth had come up to meet us. Just how this registered upon the minds of my two passengers I don't know; but they subsided into the bottom of the cockpit, and I neither saw nor heard any more of them. After we landed I understood why. The two were so utterly incapacitated they had to be carried ashore.

I never cease to marvel at how the comparatively minor hazards of flying are regarded by men who day after day face far more likely catastrophe. It reminds me of an occasion when an engineer at Rouyn engaged a crew of especially heavy and husky miners to sink a shaft; the work being pressing, he arranged to have the men flown in. I think I took in not only the first mining crew ever to be transported by air, but possibly the heaviest aggregation of humanity ever flown (to that time). Unfortunately, I contracted for this by the head. After seeing my prospective passengers, I regretted not having charged by the pound for this job. Not one of them was less than 200 pounds, and several were considerably heavier. It was obvious by the way they regarded the boat that they were not anticipating their journey with any great pleasure.

I decided to get the worst over and loaded up the four heftiest of the scared crew. The pick of the quartet, who seemed the most terror-stricken, I placed on the seat beside my own. I assured him that this was a position of much greater safety than those occupied by the others. As a matter of fact, with his colossal weight near the centre of gravity, the balance of the machine would be improved. I left him there, and, while I warmed up the engine, I observed the blond giant becoming more and more nervous. He insisted on trying to talk to me, and nothing I could do would discourage him. Finally, he became so earnestly importunate, standing up in his place and endeavouring to shout something that was impossible to hear above the roar of the engine, that I came to the conclusion it was important. I shut off the engine (which I had started only with difficulty) and asked him what was wrong. "If anything happens to me," he quavered, "tell my mother. She lives in Poland."

I informed him somewhat curtly that if anything did happen I expected to be in on the party. Then I fired up the engine once more. After an unduly long run, the craft staggered into the air, and, an hour later, I delivered a nerve-racked mass at the mine. Next day these men who had been thrown into a panic by a brief journey in the air would quite heedlessly expose themselves to the hazards of blasting, cave-ins, and other accidents—all routine to mine workers.

Laurentide employees haul a dolly-mounted HS-2L up the slipway
at their Lac-à-la-Tortue, Quebec, base. *RCAF Photo*

It was on a flight out of Rouyn that I had an experience that might
easily have been fatal. I had no passengers from the gold fields so
made up a load from odds and ends of express. Included in this
assortment were two personal packsacks and a box of ore samples.
Not long after leaving Rouyn I detected a strong aroma of burning
paper. At first I paid no attention; ground odours are much in evidence several thousand feet in the air. But the smell became so
powerful that I grew nervous. When it worsened I decided to land,
and picked out a convenient lake for the purpose.

This proved a wise course, as I soon discovered. Investigating the
box of ore samples, which had a high sulphur content, I found it in a
state of near combustion. Very shortly, in the stiff breeze, it would
have burst into flame. That would have been a pretty predicament
2,000 feet above terra firma.

The reader may be skeptical of my statement that ground smells
reach such a height; nevertheless, it is a fact. On one occasion the
odour of a skunk was so powerful up at 5,000 feet that I would almost
have sworn the unpopular civet was an unbidden passenger in the
machine.

Flying from Rouyn, I had the unique honour, as far as I know, of
carrying the first and only stowaway in the history of aeronautics (to
that time). My practice was to fly into the gold area with passengers
and express, anchor the boat out in the lake, and make the return trip
the following day. One morning, after spending the night at one of the
mines, I was paddled out in a canoe to the plane, got aboard, and fired

up. With no passengers and little express, I considered the machine lightly burdened and lost no time taking off.

As soon as I was well up in the air, however, I realized there was something wrong with the machine. It was distinctly tail heavy. Despite my best efforts, it would not fly level; the tail persisted in dragging like that of a depressed pup. I continued on, though I'll admit I was a trifle nervous and felt distinctly relieved when I finally alighted on Lake Temiscaming.

Two mechanics rowed out to help me moor the boat, and I told them my problem. They were also puzzled, since they had recently overhauled the boat. They proceeded to make an immediate inspection and soon discovered the source of the trouble. A chance look over the edge of the rear cockpit disclosed a very frightened human. We hauled him out and heard a moving story of loneliness and yearning to get out of the wilderness and see his family. But he lacked the funds. When he saw my machine, he recklessly swam out and climbed aboard. We read him a sermon on the possible results his stolen ride might have had: wrecking the aeroplane and endangering the life of the pilot as well as taking a long chance with his own carcass. But the whole affair struck us as so novel and extraordinary—and after all I had come safely out of it—that we did nothing but see him on his way to the family he was so anxious to rejoin.

MINERAL EXPLORATION BY AIR IN THE NORTHWEST

I spent the winter of 1925 seal spotting. When I returned from the sealing fleet in the spring of 1926, I joined an expedition organized by a syndicate of U.S. mining investors, for exploration in the largely unmapped regions of northern British Columbia and Yukon, as assistant pilot and mechanic. This was, to the best of my knowledge, the first mineral prospecting party to make use of the aeroplane. Col. James Scott Williams (in charge of aerial transport) and I appreciated keenly that the territory we were to fly over was isolated, largely unexplored, and inaccurately mapped. We fully understood the hazards and difficulties of such an expedition. We dug up all the available data on the region; but this was painfully fragmentary. The machine we were to fly, a Vickers Viking amphibian (G-CAEB, Napier Lion powered), arrived by rail at Prince Rupert, British Columbia, from Three Rivers, Quebec. We were there ready, with the plane assembled, to commence our trip on 1 June. There were nine people in our party.

We flew from Prince Rupert to Wrangell Island, Alaska, where we refuelled. Then we headed directly over the Coast Mountains by way of the Stikine Valley and on to Telegraph Creek, British Columbia. It was a route of exquisitely wild loveliness. We sailed over glittering glaciers, high, white-capped mountain peaks, and saw the ribbons of innumerable rivers widening below. We made a landing on the Stikine River at Telegraph Creek and established camp there to await the break-up of ice on Dease Lake, British Columbia.

On 13 June we received word that the ice had broken up. The following morning we made an early start with the entire outfit—nine men, dogs, tents, and other equipment—to fly to our main base, which was to be located at the head of Dease Lake. It was more than 70 miles from Telegraph Creek to Dease Lake over a country absolutely devoid of places to land. A forced landing would have meant failure for the expedition and, in all probability, disaster for us. Fortunately, we had no such mishap, and transported the entire miscellany to the head of Dease Lake in 45 minutes. Ours was the first machine to have ever crossed this height of land. We maintained an altitude of 3,000 feet after leaving Telegraph Creek and were only 500 feet above the water of Dease Lake—a body located 2,500 feet above

The Vickers Viking, G-CAEB, used on aerial mapping and on both northwestern Canada expeditions in which the author was involved.
C. S. Caldwell

sea level. Having ascended 3,000 feet on the takeoff, we had only to descend 500 feet to land.

Our first trip from Dease Lake was down the Dease River as far as McDames in British Columbia, where the Hudson's Bay post and a couple of mining companies were engaged in dredging operations. The 90 miles took just 50 minutes. It was impossible to convince the Indians of this, as it took them one week to make the return trip. They regarded the machine as diabolical and could not be induced to approach closer than 300 yards. They seemed to believe there was something supernatural about it. "It flies like a duck and lands like a goose," one observing Indian put it. It was the sole topic of discussion of nearby tribes, and other Indians came from miles to view it. The younger ones sat for long periods sedulously practising the roar of the engine—with their arms imitating the movements of the aircraft in banking—evidently to take back an accurate and convincing story to their own people.

While at McDames we chartered a large scow capable of carrying five tons of supplies. We hired the Indians to bring it up to Dease Lake, where it was loaded with gasoline, supplies, and equipment and sent down the river to Liard Post, located at the junction of the Dease and Liard Rivers. The trip by scow took 20 days. By air it was just under three hours.

Our arrival at Liard caused a lot of excitement. The Indians collecting on the riverbank were very much alarmed. One fellow immediately started in a panic for the bush. But, as we circled preparing to land, the shadow of the machine intercepted his flight. Terrified, he turned, deciding the river was his safest retreat, and made for it. Just as he got to the bank, however, he realized we were landing there; frenziedly he spun around and plunged back into the bush. He lay hidden there for some time before coming out to secure his pail, rifle, and dog, then set about the construction of a rude raft. That evening he set out on a 400-mile journey to Nelson to tell the chief of the fearsome thing that had arrived. I still wonder what happened to him.

While the scow was en route, we transported the prospectors by air to various locations. Considering all conditions and the difficulties a forced landing would have put us in, it was not the safest kind of flying.

The scow eventually arrived on 11 July. The two Indians who had brought it down intended to return on foot, a matter of five days and nights of travel. Since we were going back to Dease Lake and had to pass the Indian camp, we invited them to come along. At first they would not consider it at all. So strong was their reluctance and so amusing to us the attitude of these two—who had piloted the clumsy scow down all manner of rough water—that we decided to persuade them to fly if we could. It was hard work, but we finally convinced them to come on board.

We had seated them and got as far as putting helmets on their heads and cotton wool in their ears when the nose of one of the two began to bleed profusely. This was regarded as a very bad omen, and the trip was nearly off. However, after more persuasive eloquence and copious cold water poured down the back of the neck to stop the bleeding, we lost no time taking off. These two, who would have thought nothing of running rapids with a few poles lashed together with moose hide thongs, were terror-stricken. As soon as we left the water, their heads disappeared beneath the cockpit coaming and did not show again until we landed 50 minutes later. It would have taken them five days to cover the distance on foot. I will never forget their expressions as they tried to realize the distance they had come. Even greater, if possible, was the wonder and bewilderment of their friends, who had congregated on the bank to see the big bird land. They saw two of their own people emerge from the bowels of the monster.

On the evening of 5 July, we picked up the prospectors at French Creek, which runs into Dease River some 45 miles from Liard Post, and flew them to Liard. As the machine approached this post for the first time, the entire population, Hudson's Bay factor and Indians, gathered on the riverbank trying to identify the strange noise. Finally, they spotted the big bird coming straight for them from a couple of thousand feet in the air. The Hudson's Bay man, we discovered later, thought the machine was going to pass by and get lost, so he instructed the Indians to get their guns and fire a fusillade, imagining that we would hear the reports. We knew nothing about this, but a few hours after landing, we discovered considerable water in the hull of the machine and, on investigation, found a fountain shooting up just aft of the gas tanks. The fact that the hole was clean, and about the size of a 30/30 bullet, caused us to make inquiries, and we were informed about the shooting. The plane suffered little damage, though it is dif-

Old meets new: the Vickers Viking rests at its moorings as a stern-wheeler passes, probably at Fort Fitzgerald, Alberta. *C. S. Caldwell*

ficult to say what might have happened had the gas tanks been pierced.

Once we established camp at Liard Post, we carried out exploration work in many sections; the prospectors were amazed and gratified at the marvellous manner in which aerial transport speeded up such work. In one day we transported seven men, with sufficient supplies to last a month, including tents and mining equipment, a distance of 200 miles over country quite unmarked by trails and without any sign of human habitation. The first men were landed at this point and, after cooking breakfast, actually began working a couple of miles up the creek by 8:30 AM. Once the operation was complete, we flew the entire party to the main base in one day. Any other means of transportation, if not impossible, would have taken at least a year.

On the evening of 17 August, we departed from the head of Dease Lake to Frances Lake, a distance of 310 miles, about 130 miles north of the Yukon–British Columbia boundary. Here, there was no sign of human life, white or Indian, though at the old Hudson's Bay post at Frances Lake there were several deserted log cabins and a few old caches. My outstanding recollection is of the wolves, which seemed to be everywhere and kept up an incessant howling all night. They stayed out of sight, but their unquestionable presence in such numbers was unnerving.

The game in that sparsely inhabited country was unbelievable. The lakes and rivers teemed with the finest fish, which, in addition to being an important part of our diet, furnished us with superb sport. Flying over the country, we often saw moose, sometimes as many as a dozen at a time, standing at the shore of small lakes. As we sailed along close to the high-ridged mountains, caribou, goat, grizzly bear, and other animals were visible at almost any time. On one occasion I saw a pure white moose. Another time, when I was fishing, I saw a huge moose and called to Col. Williams to come and have a look—he was casting farther upstream. When he did not come immediately, I went to see why. He had been intercepted by a bear with two cubs and was busy bluffing them. Luckily his bluff wasn't called.

The last place we visited before coming out was the fabled "Tropical Valley." While at Liard Post, we had heard vague accounts of the existence of this mysterious valley. We had also heard about a prospector and trapper, Tom Smith, and his daughter who had wandered overland from Yukon into that region two years before and had not been seen or heard from since. We decided to make the trip to see if the valley really existed and if possible find some trace of the missing pair.

We knew that the valley was approximately 200 miles from the post and accordingly took off and followed the course of the river, which

Deep in the northern British Columbia wilderness, Jack Caldwell prepares a meal for the Viking's crew and passengers. *C. S. Caldwell*

is punctuated with terrifying rapids. We had flown what we judged to be the approximate distance when we reached a region of peculiar looking lakes. There was, perhaps, nothing unusual about them, but they were striking to us because there were no similar bodies of water in the country for some considerable distance in any direction. Deciding that this must be the place, we landed on the river and tied up. With the machine secure we headed out to scout.

Almost immediately we struck a trail, faint and apparently unused for a long while, but unmistakably a trail. We followed it for about 400 yards from the river and then, to our surprise, encountered a board nailed across a tree. Upon a piece of wood painted in crude characters were the words "A.B.C. Code" followed by a jumble of figures and signed "Tom Smith." It baffled us for a while, but the combined wits of the party finally deciphered the words "Message in bottle at foot of tree."

We dug at the bottom of the tree and just below the surface struck a bottle that contained two pieces of paper covered with very legible handwriting. I cannot remember the exact wording, but it started off, "To any white man finding this message" and went on to the effect that by following the trail back for two miles you would find a cabin near the first hot springs and a short distance beyond, a garden. "We have left planted potatoes and onions." By further following the trail, the message continued, the finder would come to a large hot spring overlooking a meadow "in which moose are to be found every morning and evening during the summer months." The message, signed by Tom Smith, concluded by saying he and his daughter had been living there for two years and had seen no white men and were leaving for Fort Simpson, which, "barring trouble with the Indians, we expect to reach in the spring."

We progressed up the trail, the atmosphere becoming more torrid and languorous as we advanced. Shortly, we reached the first hot spring and just beyond that found the deserted cabin. A little farther along was the garden, rank and overgrown. From here we started out to explore a limited area of the valley, which appeared to be about 10 miles square in extent.

Hot springs sprang from the ground all over. Some of them just bubbled up and ran away. Others formed the pools of varying dimensions we had seen from the air. The rich and luxuriant growth and foliage were distinctly suggestive of a tropical region. Ferns grew to an enormous height and size, and vines spread all ways in a tangled mesh. Berries of many kinds grew in profusion and were of extraordinary size. Flattened patches indicated where bears came to dine. Large patches of purple violets of a size and beauty I had never before known grew about the lakes. Much of the growth was unfamiliar to me. I had not seen anything as lavish in other parts of Canada. I was brought up in Alabama, and nothing has ever so reminded me of that southern state.

We stayed two days and then flew back to Liard. When we reported what we had found to the Mounted Police at Fort Simpson, they told us the rest of the story of Tom Smith and his daughter. The Smith's canoe had capsized in the rapids as they were coming down the river. The girl had been swept away and came to rest, unconscious, on a sandbar. When she came to her senses, there was no trace of her elderly father. She had made her way to Fort Simpson and was now working for the Hudson's Bay Company.

The summer's work complete, we returned to Liard Post and left there on the morning of 28 August, landing at noon of the same day at Wrangell. From there we proceeded to our starting point, Prince Rupert, arriving on 31 August, having covered the distance of over 300 miles in a little over four hours. Thus was successfully terminated the first aerial mining exploration trip in the history of aeronautics, one of the most single-handed and isolated of aircraft operations ever—carried out without a hitch.

After one more winter with the sealing fleet, I returned just in time to join another mining expedition, this time instituted by an Alberta syndicate into the Northwest Territories. The party consisted of five, including two mining engineers; again I was assistant pilot and mechanic. We shipped the same Vickers Viking amphibian to Edmonton thence to Lac la Biche in northern Alberta, where it was assembled. From Lac la Biche we flew to Fort Fitzgerald, Alberta, and from there we set out for a point 400 miles northeast where we planned to establish our main base. Unable to make this in one flight, we accomplished it in stages with two intermediate camps and caches about 150 miles apart. We stayed from one to two weeks at each, flying the mining engineers out on various radii to prospect.

It was a novel enterprise and not without its flying hazards. North of Fort Fitzgerald there were none but the Indians, and this had apparently been the case for 25 years or longer. The meat drying racks we found were at least that old. Farther north we entered the region that has been aptly termed "The Blind Spot of Canada," an immense, desolate stretch of badlands, bald rolling barrens, with periodic scrub timber. No maps existed of the area, and we were only able to determine our position from the sun and the stars. We passed over literally hundreds of lakes and rivers as yet uncharted. Flying from one cache to another, these were our only guides. One had to fix the shape of certain bodies en route in the mind—the area was singularly devoid of characteristic features—or else make one's own crude maps. This was both a strain and a worry. If we missed locating a cache, we would have to make a forced landing, with fuel running out. That would have been a pretty predicament. Today, it is generally believed that one cannot find any place in Canada where the white man has not been; but with the Viking, at that time, it was comparatively easy. We passed over thousands and thousands of miles of unsurveyed territory and made landings and established camps in the most absolute wilderness. Had anything gone seriously wrong with our aircraft, it is quite possible that no one would ever have learned of our plight.

Having established our main base, we spent the entire summer transporting supplies from the earlier two caches and in making radial prospecting trips, all accomplished without mishaps of any kind. The engineers seemed entirely satisfied with their work, and we brought back heaps of ore samples. While the care of the aeroplane and the maintenance of an adequate supply of gasoline kept us pretty busy, there was still time for entertainment. Except for the mosquitoes, which were numerous and voracious, there were no drawbacks to this life of isolation.

The resources of this country, even the tithe of them that is known, are immense. Previously unknown lakes teem with the most edible fish and will someday be of high sporting as well as commercial value. As we ventured farther north we encountered herds of thousands of caribou. Seen from 2,000 feet above, the myriad trails of these animals spread like a mesh over the bleak surface of the Barrens. Caribou constitute the principal sustenance of the Indians and of such white trappers as penetrate the area in search of furs.

On our way out, with the summer's work completed, we were

The Fokker Super Universal, CF-AEX, in which Jack Caldwell lost his life after striking hydro wires above the St. Lawrence River on 20 June 1929. *H. C. W. Smith via K. M. Molson*

forced to make a landing on a lake about 100 miles from Fort Fitzgerald. As we waited for the weather to clear, two Indians of the Caribou Eaters tribe paddled out to inquire partly in English and partly in French whence we had come. From the Barren Lands, we replied, or the "Land of No Sticks," as they call the region. Immediately they were eager to know if we had seen any caribou, and, when we replied in the affirmative, they were greatly pleased—this meant plenty of meat for the winter.

One of them was most anxious to send a note to some friends at Fort Fitzgerald and asked if we would deliver it. On hearing that we would be only too glad, he produced a notebook and, with much laborious effort, which brought beads of perspiration to his brow, he penned a queer geometric note. After its slow completion he handed me a dollar bill. I declined it and asked him how long it would take him to go to Fort Fitzgerald himself, and he replied, "Long time. Lots of portages. Maybe one month." When I told them we could get to the fort in under an hour, the pair was dumbfounded and incredulous. "Far, far, very far; no not far," one of them said, evidently intending to convey that what was a long distance to them was little to us.

The next day was sufficiently clear to permit us to resume our way, and we reached Fort Fitzgerald. From there we flew to the government aerodrome at High River, Alberta, passing over Edmonton en route. In all we covered 7,000 miles without trouble or mishap. Our staunch little amphibian presently hibernates while I taste the things of civilization again, and savour a little time to reflect. While there certainly seems to have been a lot crowded into my last four years, each experience merely whets the appetite. Soon the spring will be around again with the Red Gods beckoning to the romantic and adventurous life of a freelance pilot.

Flying on Civil
Operations with the
Prewar RCAF

J. D. (JACK) HUNTER

Jack Hunter died in October 1998, 20 years after retiring from the Department of Transport. The following account appeared in the Spring 1991 CAHS *Journal*. Jack's aviation career began in 1928 when he joined the RCAF at the urging of a young flight lieutenant friend. At that time, pilot trainees required a university degree. With only a high school diploma, Jack opted to train as a carpenter/rigger—a logical trade in an era of wood-and-fabric aircraft. Subsequently, he remustered (applied for training) as a metal fitter and then an Aero engine fitter. In 1931 he began training as a pilot at Camp Borden, Ontario.

During the 1920s and 1930s, only the Ontario and Manitoba Governments operated their own air services. Under contract, the RCAF provided forestry and fisheries patrols and other "civil operations" for the remaining provinces. As well, the air force undertook the aerial mapping of much of Canada, although mapping of the far northern regions would not be completed until the 1950s, when the work was completed by a combination of the RCAF and commercial operators. This work constituted the principal involvement of the prewar air force. Its fighting capability was limited to a dozen 1920s-vintage Siskin biplane fighters (which put on spectacular aerobatic demonstrations at airshows across Canada) and a few Atlas and Wapiti army cooperation aircraft of comparable age.

Much of Jack Hunter's subsequent flying was in the Canadian

Vickers Vancouver, a twin-engine biplane flying boat with a metal hull and wood-and-fabric wings and tail surface. Dimensionally it was similar to the modern de Havilland DHC-3 Otter, and the Vancouver's two Whirlwind engines provided about the same power as the Otter's Wasp. Weights too were comparable for the two aircraft. However, with its drag-inducing rigging, the Vancouver's performance was far below that of the Otter. Today this attractive aeroplane is remembered only in photographs.

Jack Hunter's sense of humour stood him in good stead throughout a long flying career, and in later years he was known and appreciated as an amusing raconteur. He begins his story following the presentation of his RCAF pilot's wings.

Upon our graduation from Camp Borden, my former classmates and I were posted to RCAF Station, Jericho Beach, Vancouver, for seaplane training. The conversion course usually lasted about six weeks, but, as so often happened, the air force was broke, and there was no money to transfer us back east. That was no hardship as far as we were concerned, and we managed to stay at Jericho Beach for a full year.

Our commanding officer at Jericho was Squadron Leader Earl McLeod, with Flt. Lt. A. deNiverville as adjutant. Instructors were Flt. Lts. deNiverville, Bennett, Holmes, and Mawdesley. We trained on Gipsy Moths on floats, Vedettes, Vancouvers, and the sole Consolidated Courier at the base. We also had the one and only Vickers Vista, a tiny, single-seat flying boat that we used exclusively for taxiing practice. After becoming reasonably proficient on these types, we were assigned to various operations. We often went out on fisheries patrol to check on the amount of net the fishermen were using in Howe Sound and up the Straits of Georgia. One day I landed beside a Japanese fisherman who was using about twice the legal footage limit. I signalled to him to approach the aircraft. Instead, he started up and charged at me full bore. I pulled a Very pistol out of my map case and let one fly at him. The Very light bounced off his deck, and you never saw a boat go into reverse as fast in your life. He didn't want any part of that, and I can't say I blame him. I copied down his boat licence number and turned it in to the Fisheries people for further action.

Author Jack Hunter (left) in flying gear (sidcot suits) with
fellow pilot trainee Norv ("Molly") Small. *J. Hunter*

We also used to go on narcotic patrols. We would pick up an RCMP observer and head about 20 miles out to sea to meet one of the *Empress* ships coming from the Orient. Then we'd circle the ship all the way into Victoria. Some of the crew had a habit of heaving sealed tins, presumably containing drugs, overboard. A fishing boat would pick up the tins and drop them somewhere along Vancouver Island's west coast. The patrol was quite successful. I never spotted any floating tins while we were in the area.

Another time I was assigned to a special patrol on the west coast of the island when the U.S. Coast Guard spotted a schooner off the Washington coast. The ship was apparently carrying a number of Chinese who planned on landing illegally in some secluded cove along the U.S. coastline. The coast guard tailed them for several days and then passed the word to our Immigration people when the schooner entered Canadian waters. They asked us to set up a patrol to search out the coves and inlets on the west coast of the island.

It was late fall, and the weather in that area was certainly not the best. We used a Vancouver to search the coast from Esquimalt to Barkley Sound through rain and fog patches. A Royal Canadian Navy destroyer stood by with our fuel. We pulled up to the destroyer's stern and made fast with a bow and two wing lines. I remained in the cockpit of the aircraft to assist in the refuelling. When a sudden squall came down the mountains, the destroyer started to drag anchor. This was no place for the skipper, so he took off—with us in tow. The wind was so strong that by easing back on the control column I could get airborne behind the destroyer. We were in tow from shortly after noon until nine that night when we finally found shelter at Port Renfrew. By that time, I was soaked to the skin, and so cold I had to have assistance boarding the ship. Down in the wardroom, a glass of Navy rum helped stop my shaking but made it all the more difficult to get into my bed—a hammock—for the night.

We searched for two more days, but because of fog and rain and general poor visibility, we had to call it off and return to base. Surface craft continued the search but did not locate the schooner. I never did hear whether or not the Chinese landed on our coast.

In those days, the RCAF had no radios in their aircraft so when we took off on a cross-country flight of any distance we carried along a wicker basket of carrier pigeons. If we ran into trouble, we would scribble a note, attach it to a pigeon's leg, and release the bird to carry the news back to the base. We had a large pigeon loft at Jericho and often took birds out and released them for practice and exercise.

One day I was to fly a passenger to Esquimalt on Vancouver Island. Flt. Lt. deNiverville told me to take along a few pigeons and release them before returning to Jericho. The regular pigeoneer was not on the station that day so another lad, who claimed he knew all about the pigeons, packed a basket and off we went. When I released the birds at Esquimalt, they seemed reluctant to leave the aircraft. I had to shoo them away two or three times before they climbed, circled, and disappeared from view behind the mountains.

When I got back to Jericho, I could see our CO, Squadron Leader Earl McLeod, pacing the beach, waiting for me to taxi the Vedette to shore. Something was wrong. I no sooner beached the aircraft then McLeod said, "You didn't release the pigeons, did you, Sergeant?" "Oh my, my, my!" was all I could say. It turned out that these particular birds were from the King's loft in Britain and were for breeding purposes only—they had never flown in this country. We waited around the rest of that day, but they never did come home.

As all good things come to an end, so did my sojourn at Vancouver. On 23 May 1932, I took up a new post at Ottawa Air Station. After settling in I found that my assignment was to take part in the experimental airmail flights from Red Bay, on the coast of Labrador, to Ottawa. That was also the summer that Ottawa hosted the Imperial Economic Conference.

The *Empress of Britain* normally crossed the Atlantic from Southampton to Belle Isle, Quebec, in about three days and 16 hours, then took another 29 hours to reach Quebec City. This meant that mail carried by the liner did not arrive in Montreal until 48 hours after the *Empress* had passed the Strait of Belle Isle. If the mail was transferred from the liner to an aircraft in the strait, it could be delivered to Montreal at least 24 hours in advance of its normal arrival, and, similarly, outbound mail could be dispatched from Montreal by air 24 hours after the vessel sailed.

Under the proposed experimental air service, a naval minesweeper would meet the *Empress* as it entered the strait at dawn on its fourth day out, transfer approximately 800 pounds of mail, and deliver it to two Bellanca seaplanes waiting in the sheltered waters of Red Bay. The

RCAF Bellanca Pacemakers, G-CYUX and G-CYUZ, at Havre St. Pierre, Quebec, figure prominently in Jack Hunter's story. They were used in conjunction with the Canadian Vickers Vancouvers that he flew. *RCAF RE-18043*

Bellancas would then fly the mail to Havre St. Pierre on the north shore of the Gulf of St. Lawrence, some 600 miles below Montreal. There it would be transferred to two Vancouver flying boats and flown across the gulf to Rimouski, Quebec. Transferred again to a Fairchild 71 landplane, the mail would be flown to Montreal and thence to Ottawa. In this way, it would arrive for the conference within four days of dispatch from Southampton.

The Belle Isle Detachment was under the command of Squadron Leader R. S. Grandy, who was based at Rimouski. At Red Bay, with the Bellancas, were Flt. Lts. N. C. Ogilvie-Forbes and Sgt. F. J. Ewart. At Havre St. Pierre were the Vancouvers, with Flt. Lts. A. deNiverville and F. J. Mawdesley, Sgt. H. Bryan, and myself. Flt. Lt. D. A. Harding, Flying Officer E. A. McNab, and Sgt. Bowker handled the run from Rimouski by landplane.

An earlier trial flight had to be cancelled due to bad weather, but on 12 July the first scheduled flight between St. Hubert and Red Bay took place with 317 pounds of mail. That flight between Rimouski

A brand new Canadian Vickers Vancouver prior to the application of RCAF markings and identification. In the background is a Vedette, also unmarked. The Vancouver was comparable in size and weight to the modern de Havilland Otter. *RCAF HC 2057*

and Havre St. Pierre was not exactly without incident, and with due apologies to the late Dr. William Henry Drummond, I took time afterwards to record the high spots of that stage of the trip. If you will bear with me, I will offer my first and last attempt at verse.

Par Avion, July 12, 1932

I fix up dis here poem
For hexplain to my girl
Of de flight we 'ave on Juillet 12
De firs' run of de mail.

We get report de bonne heure dat morn
De wedder she's look fair.
At six heures sharp, we den decide
We're gonna take de air.

We warm dos two engine up
Make sure dey give de revs,
Pile in de mail, give her de gun
We're off before I'm dress.

Dinny, he's fly de firs' demie heure
Den he's pass de stick to me.
I set my course for Cote du Nord
An' strike for open sea.

We fly deux heures an' den, by gar
Dat fog she's close in fas.
We look ahead, we look behind
An' wonder how long she's las'.

At las' we hit de big fog bank
Jus' off of Manitou.
She's come so quick, by gar, we're trap
Ever't'ing she's gone from view.

We give dos engine beaucoup gun
An' climb toute suite like hell.
We don' know what's ahead on lan'
But la mer we know ver' well.

We circle 'round one, two, t'ree time
We look for hole below.
An t'rough one ver' t'in spot
By gar, we see de l'eau.

We make de landing at dis place
An' drif' for hour or more.
We can't see no'ting but taxi nord
An' by n' by strike de shore.

We wait here for de fog to lif'
We stay a full hour more.
Au bout de quelque temps she's fine
An' away we go encore.

We pass by Sheldrake an' T'under Bay
We t'ink we make St. Pierre.
But when we get to Magpie
Dos fog she's t'ick dere.

We 'ave to make de turn aroun'
An' land at T'under Bay.
We wait two hour for report
Dat de fog, she's clear away.

We take off den at cinq heures sharp
An' fly to Havre St. Pierre.
It sure look good w'en we get down
To see dose old gang dere.

We put de machine on to de buoy
An' make fas' for de night.
Did someone say, "You hungry, boys?"
By gar, we are, dat's right.

We go ver' quick chez Mme. Laundry
An' 'ave de juicy steak.
By gar, she's good for eat again
My stomach she's near break.

An' now dat day she's gone at las'
We're tired like a dog.
We t'ink dat machine never was
Make to fly t'rough fog.

So you see, my dear, dos letter I send
She's got de historie.
Of de fir's mail run on Juillet 12
An' was carried part way by me.

That, by the way, wasn't the only time we were down in the gulf.
Fortunately, the second time we weren't carrying mail but were on a
transportation flight. We were outbound from Havre St. Pierre to
Rimouski, and about halfway across the gulf, or about 150 miles from
Rimouski, we suddenly lost oil pressure on the port engine. There was
a fair sea running, and it was not the best of times to be wandering

around on the wing, removing engine cowlings and the like. When I loosened the nut on the oil pressure relief valve, the valve stuck, then popped out. Dinny bobbled it, and it bounced once on the wing then sank in umpteen fathoms of water. That was a pretty spot to be in, and of course we had no spare with us. I used the relief valve from the starboard engine as a pattern, and, using a quarter-inch bolt from my tool kit, and with only a hand clamp and file, worked from 10 that morning until four in the afternoon to fashion a new valve. After a few minor adjustments, the pressure on both engines held and we were off to Rimouski. When that particular engine was overhauled later on, it came back from the manufacturer, Canadian Wright, with my homemade relief valve still intact!

During the latter part of September that year, I was posted to Shediac, New Brunswick, to replace one of our pilots who had lost a couple of fingers to the propeller of a wind-driven generator. The Shediac detachment was one of the four in the Maritimes engaged in RCMP preventative patrols. In other words, chasing rumrunners. We were flying Fairchild 71s on floats and carried an RCMP constable as observer. Our patrols lasted up to six hours, and we covered approximately 500 miles of coastline. The rum-running schooners would load up in the French islands of St. Pierre and Miquelon and cruise just beyond our coastal waters, where they would rendezvous with high-powered speed boats. These fast boats would make a quick run

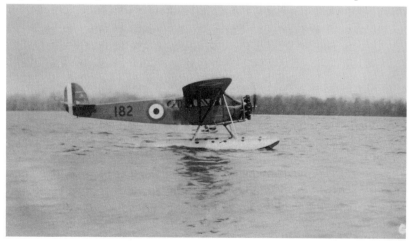

Jack Hunter flew the Fairchild 71, typified by RCAF 182, on smuggling suppression in the early 1930s. *J. Beilby*

for shore with their illegal cargo. The RCMP had cutters patrolling the area that were in contact with us as we pounded out Morse code by CW wireless to indicate the positions of the schooners and their speed boats. It was all very interesting, and on several occasions we witnessed exciting chases, sometimes with a goodly exchange of fire.

The RCMP caught so many of these rumrunners that the local jails were full to capacity. It got to the point that when a rumrunner was convicted and sentenced, he had to wait for a vacancy, and in many cases went right back to his old trade until there was room in jail. They swore up and down that if ever they caught us down at sea, they would put us under. I, for one, believed every word they said. It was a good incentive to stay up.

One memorable incident took place in late evening when I was on my way back from Charlottetown, Prince Edward Island, to Shediac. It didn't involve a rumrunner; this was a lobsterman. I was right down on the water, beating my way against a strong headwind, when I flew low over his boat. As I approached him, one of the two aboard held up something that looked like an oar. Then I saw flame come out the end, and some shot pellets ripped through the fuselage just behind my seat. Instinctively, I pulled back on the stick. It suddenly dawned on me that he had actually taken a shot at my aeroplane. I turned down after him, but the old Fairchild squashed a bit coming out of the dive, and I hit both gunnels of his boat with the keels of my floats—he was flat against the floor boards by this time. It was dark by the time I got back to Shediac, and I had to line up by a big kerosene-burning lamp at the end of the dock. From there I had to feel my way down—landing toward the beach so I could ground the aircraft when I reached the end of my landing run. The damage to the keels of the floats was not excessive, and local repairs took care of it. The operation closed off the season at the end of October, and I returned to my base at Ottawa Air Station.

With Test and Development Flight department of the RCAF, we were doing some test flying on a device that would help pilots land in fog or when smoke greatly reduced visibility. It was a crude system, but it worked. A Vedette flying boat trailed a weighted cable of prede-termined length behind it; when the weight struck the water, it cut in a sensitive altimeter that was graduated from 80 feet to zero. It was a large-faced instrument with widely spaced graduations. By this

The graceful Canadian-designed and -built Canadian Vickers Vedette was a light RCAF patrol aircraft occasionally flown by Jack Hunter. *RCAF A-737-192*

means a long power-on approach could be made down to about five feet above the water, then the throttle was closed and the hull would settle nicely onto the surface.

In June 1934, I was based at Hamilton, Ontario, on a photographic mapping mission covering the Niagara peninsula. In this exercise I used a Bellanca equipped with autopilot. In mid-season I returned to Ottawa to carry out some tests on long-range infrared photography. With a special mount and camera, I operated from 18,000 feet over Ottawa and photographed Montreal through heavy haze. The Bellanca was operating at close to its service ceiling, and we flew without oxygen, a tiring experience, I can tell you. We completed photo operations of the Niagara peninsula in late November, and I returned to base in Ottawa.

I spent a great part of the summer of 1936 instructing on seaplanes and flying boats at Ottawa Air Station. Again the air force was short of money and the members of the ab initio class, who had completed their training at Camp Borden, were assigned to Ottawa Air Station for seaplane work instead of going out to Vancouver. This training was all right, but it took place in sheltered waters, which meant we were unable to provide trainees with the experience that only the open sea can offer.

One day I went to Montreal to pick up a newly overhauled Bellanca from Canadian Vickers. I picked up the test flight documents from the pilot at Vickers, then taxied out to takeoff position, checked the controls as far as I could see, set the stabilizer to the normal takeoff position, and opened the throttle. The aircraft was tail heavy on the first part of the takeoff run, and I wound forward on the stabilizer control to correct this situation. The aircraft, being lightly loaded, leaped off the ground and started to climb like a homesick angel, just on the point of stall. It suddenly dawned on me where the trouble was, and I frantically wound the stabilizer in the opposite direction. This lowered the nose and let me gain flying speed. I circled the airport, landed, then made a beeline for the test pilot's hut. I asked the test pilot when he had last flown the aircraft.

"Oh. Just a couple days ago," he said. "That's one beautiful machine to fly."

I asked him if anyone had worked on it since that time. He indicated that the aeroplane had been put away in the hangar right after the test. At this point I blew my top and asked him if he always flew the aircraft with the stabilizer rigged backwards. Of course he denied this, so I took him over to the chief inspector and verified it. No one could understand how it had happened. But it proved one thing—the test pilot had never had that machine off the ground.

In December, we took delivery of three Northrop Deltas to be used for aerial photography. It was the first low-wing monoplane I had ever flown and, for that period, was a high-performance aircraft. Being attached to Test and Development Flight, it was my job to carry out the camera tests. We took off on the first test and pulled the camera plugs at 10,000 feet and proceeded to do a practice run on a photographic line. It wasn't long before I started to feel a bit lightheaded. I told my camera operator to replace the plugs and that I was going down. I made three passes at the aerodrome before I could get the thing on the ground, then staggered out of the aircraft and reported to the CO that when the camera plugs were pulled a heavy concentration of fumes entered the aircraft.

Max Kuhring and several of his cohorts from the National Research Lab came down with a lot of test equipment, and the next

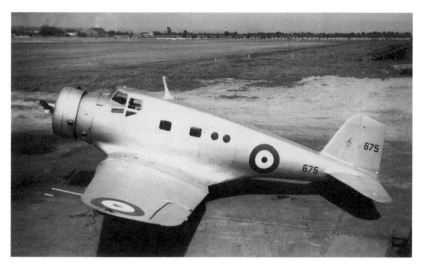

Northrop Delta, RCAF 675, similar to the machine flown by Jack
Hunter, was a cleanly designed machine and a good performer.
Via D. E. Anderson

day we donned gas masks and again went up to altitude and removed
the camera plugs. After about one-half hour we landed, and Max took
his air samples to the lab. The tests showed there was a heavy enough
concentration of carbon monoxide to kill in a half-hour flight. After
that I put in about 25 hours flying with a gas mask, trying different
methods to build up a positive pressure in the cabin. These attempts
were unsuccessful, however, and the Deltas were pulled out of service
for modification to the exhaust system to permit the collector to
exhaust over the wing instead of along the bottom of the fuselage.
This modification corrected the problem, and we were never exposed
to exhaust fumes from then on. As I remember, though, it sure was
uncomfortable flying with a gas mask in cold weather. Every time you
exhaled, the goggles would fog up. A half-hour of that at a time was
more than enough.

In the spring of 1937 I was slated for a photo operation with the
Deltas to do some mapping out on the west coast of Vancouver
Island. We were to fly the aircraft out on floats. It promised to be a
very interesting summer.

Jack Hunter never did get to fly Deltas on the west coast mapping operation. At the behest of his former CO, "Tuddy" Tudhope, who had earlier resigned from the RCAF and accepted a senior position with the Department of Transport, Jack also resigned and joined Tuddy as his assistant. He remained with the department for the next 33 years, piloting the Lockheed 12, CF-CCT, which is now in the National Aviation Museum, for much of that time. When he retired in 1970 he was superintendent of maintenance for the department's extensive fleet of helicopters and fixed-wing aircraft.

A Pilot's Wife

Bush Flying from the Distaff Side

LUCILLE BURTON

Lucille Terwin took her first aeroplane ride in May 1927 at Daytona Beach, Florida. Three years later, she married Ed Burton, an airmail pilot with Canadian Airways Ltd. At the time she was a recent university graduate and budding journalist. Her training as a reporter may have fostered her love of adventure and her natural curiosity, which would stand her in such good stead in the future. Ten years later she wrote the following insightful account at Perron, Quebec, where her husband had been based flying a Curtiss Robin for Amos Airways Ltd. When cutbacks left him without a flying job, he managed a mine's general store to support his wife and young son, and studied the mining industry.

Edward Cherry (Ed) Burton had earned his wings in the Royal Flying Corps (RFC) during World War I. Training on Curtiss JN-4s in Ontario and Texas, he had demonstrated talent and had been obliged to remain in Canada as a flying instructor rather than being sent to France. Upon his release from the RAF, Burton found it impossible to obtain a flying job as there was an overabundance of trained pilots. After stints as a bookkeeper in Toronto and then a clerk at an Ontario bush (lumbering) camp, he learned that the Ontario Government required pilots for its new air service. In June 1924, after a short refresher course, he received Canadian Civil Licence No. 196. Over a long and varied career he would fly for the Ontario Provincial Air Service (OPAS), the Toronto Flying Club, National Air Transport of Toronto, Canadian Airways Ltd., the London (Ontario) Flying Club, the Ottawa Flying Club, Dominion Skyways, and Leavens Brothers Air Service of Toronto. As well he flew two lengthy mineral explorations into Yukon and the Northwest Territories. His civil flying concluded with the Ontario Department of Lands and Forests (as the old OPAS had been renamed), where it had begun.

For practical reasons, wives of bush pilots seldom flew with their husbands. Apart from the fact that paying passengers and cargo had priority, there were the demands of maintaining a home and caring for children. And many wives saw riding in a bumpy, draughty aeroplane as a trying experience to be tolerated only when absolutely necessary. "Cile" Burton was an exception. Perhaps her early training as a journalist allowed her to accept, and even enjoy, flying. Lyn Leigh, wife of the well-known Z. L. ("Lewie") Leigh, also did a substantial amount of flying with her husband. Mrs. Burton's story (written c. 1940) is unusual because it is one of only two accounts known by this writer to have been written by the wife of a Canadian bush pilot. Connie Dickins, wife of the famous C. D. ("Punch") Dickins, wrote the other—a booklet entitled "I Married a Bush Pilot." Lucille Burton seems to have accompanied her husband more often than most bush pilots' wives. These experiences undoubtedly gave her a special understanding of the daily challenges her husband faced. Mrs. Burton died in 1996 in Thunder Bay, Ontario, at the age of 90.

It has been fun and scares and thrills pell-mell, these 10 years of being married to an aeroplane pilot. Sombre days of uncertainty, sharp moments of fear, and then again rollicking times when I have felt the tremendous fascination of the game, enjoying flights and seeing scenery denied to earthbound folks below. Through it all has been the grand adventure of having a part, if only to stand by and ram the powder in the muskets as it were, in a movement that has pushed back a nation's frontier by hundreds of miles and advanced its progress to an incalculable degree.

Taking history by the forelock and wresting it from the future to the present is what aviation has done for Canada. For more than 20 years, flying has been my husband's career; from the devil-may-care war times when any landing you could walk away from was considered a good landing, to the scientifically controlled enterprise it is today. During this time he has been engaged in various ventures: early commercial flying when unsuitable war-type planes were used, haphazard barnstorming, instructing, piloting scheduled air mail, and now, perhaps most interesting of all, bush flying in Canada's rich and tempting north country.

For a good part of this time I have been beside him, sometimes holding his coat and consenting, at others fiercely rebelling at the hardships, struggles, and risks. A Chinese philosopher has said that the pleasure of scratching an itchy spot more than compensates for the irritation caused by the eczema. The thrills I have experienced in receiving my pilot-husband back safe from danger and near-death have been wonderful, my pride in his achievement, great. But dare I say they are counterbalanced by the agony of those vigils when I waited for his arrival or word of his safety?

Lucille Burton bids her son Ted and husband Ed goodbye as they prepare to leave the Ontario Provincial Air Service's base at Sault Ste. Marie, Ontario, for Orient Bay, another OPAS base on the south shore of Lake Nipigon, Ontario. Their aircraft is the OPAS's only Vickers Vedette. Mrs. Burton would join them later, by train. *E. Burton*

One of the biggest compensations of being married to this game, or being married to a man who is married to it, is found in the interesting people with whom one comes in contact. There have been Indians and those in charge of Indian affairs; student fliers, from beardless youths to retired financiers, alike eager to woo the will-o'-the-wisp of the heavens; yokels and their girls at country fairs, the former with their nonchalance belied by their tense clutch upon the cowling, the latter frankly nervous and giggly; surveyors with their parties of men running lines through hitherto uncharted lands; engineers scouting for hydroelectric sites, locating railroads, or performing some of the other necessaries that go with the advance of "civilization"; and the prospectors, who are such delightful company, spinning their endless yarns.

Prospectors usually travel in pairs. One chats with them as they load their equipment into the aeroplane, the pots and pans, axes, food, packsacks, and tent. When their canoe has been lashed to the float struts, they squeeze on top of the load and away they go over the tree tops, 50, 100, or 200 miles into the wilderness in quest of the find that will make them wealthy overnight, optimistic always. As the pilot leaves them on the shore of some distant unnamed lake, they call out, "Don't forget to call for us eight weeks from today."

One prospector told me he had just returned from a district where there was a gold rush. He said there were so many fellows in the bush staking, "I was lucky not to have a leg cut off for a staking post." He added that he was going north again while he had two legs under him.

It was a decided shock to have someone write and ask, "Do you keep your ear glued to the short-wave receiver when your husband is flying?" And, "Are you not terrified when there are periods of lapsed communications?"

In most cases, communication "lapses" from the moment the aeroplane soars away from the home base until the heavens yield it back again, gently lowering it to the spot from which it was first lured. There are communications of a sort. If engine trouble is encountered in the bush, it is sometimes possible for the pilot to send word to the base by an Indian who can perhaps cover the distance in a day's journey that the aeroplane made in a 20-minute flight.

There was no communication between the base and the trading post towards which my husband was bound late one winter's day.

The Canadian Airways Stearman 4EM, Junior Speedmail, CF-AMC, which Ed Burton was forced to abandon by parachute in December 1931 when he ran out of fuel in dense fog near Dunnville, Ontario, was a sister ship to CF-AMB, seen here at Rockcliffe (Ottawa). *J. Bielby*

About an hour after he had left, a blizzard closed in and shortly afterwards, darkness.

"Lady, I'm afraid you won't have your husband home tonight" was the message I received from the airport.

"I know it," I replied miserably. "He has hardly had time to reach the camp, and there will be no daylight today."

The following day, it was blowing a gale and there was no sign of the awaited aeroplane. Perhaps, we speculated, the wind was too strong for a takeoff as a light craft had been used and the site of the post was unfavourable. You have to have some point upon which to build your hopes. At any rate, we believed he was ensconced in a comfortable camp. The morning of the third day was fine but brought no sign or word concerning the missing pilot.

In mid-afternoon he arrived, face blackened from two nights spent hunched over an open bush fire, clothes sooty and bedraggled, his boots unlaced—for 60 hours. I hurried to prepare a cup of hot beef extract.

"No thanks," he said quickly. "I've lived on beef extract for three days."

He explained the hold-up. When he was three-quarters of the way to his destination, the blinding snow had come, and it was necessary to effect a landing while he could still see.

He came down on a frozen lake and taxied up to the shelter of the bush. There was no tent in the aeroplane so he used the engine cover to provide some protection against the elements, and green jack pine afforded a fire. That night, it rained and froze alternately. The next morning, the wings and other surfaces of the aeroplane were coated with ice. The pilot's first job was to brush off the ice, swishing at it with spruce branches. The manicure was finally completed, but the gale was too strong for a takeoff. Not until darkness fell again did he give up hope of getting away. As he had had no thought of spending another night in the bush, it was necessary for him to stumble about in the dark, collecting firewood and making his primitive camp. On the next day, after several hours, the wind dropped. He managed a successful takeoff, and the aeroplane was home again in a short time. It was well we had no communication for I had imagined him safe in a warm camp all the time.

It has been just as well on other occasions, too, that I did not have in my mind the true picture of events that had transpired. The first word I received from my husband one night when he was flying in an impenetrable slate fog was that he was "safely down" with "a bit of trouble." From reporters, I learned that he had taken to his parachute when he could find no hole in the floor of fog that blanketed the mail routes from Toronto to Detroit. On another occasion when the skis of his aeroplane had broken through soft ice, he appeared home safe and sound and smiling before I heard the story.

From feeling numerous anxieties when there was cause for none, and again suspecting no trouble when there were serious things amiss, I believe my judgement has become warped in regard to the dangers of flying. Sometimes when my husband is away, unreported, and I attribute the delay to some feasible cause, I start to worry only when I sense that my friends are sitting up nights with me as it were, God bless them. Casual, guarded inquiries are made as to when I expect my hubby home. Conversation in my presence is light and airy, invitations are profuse, and I realize there is some speculation and worry concerning his predicament. Of course, I begin to worry too.

The best antidote for the misgivings of a flier's wife is for her to take a trip with her husband. Some companies encourage pilots' wives to fly in order that they may share the thrill that comes with flying and sympathize with the fascination it has for their husbands. I would not change for all the jewels of Opar the memory of some of the flights I have had with my husband at the stick. Circling over Detroit by night, its thoroughfares lined with strings of brightest gems. Hedgehopping over the fields and environs of Ottawa. Cruising over the ravines and pleasant countryside of Toronto. Speeding over the wooded stretches of Maine and New Brunswick, the mountain peaks undwarfed even by our height, the land below a vast relief map, snow-covered.

And now, north country flying, with something new in every trip. I have often made such flights perched on a load of freight, frequently with a canoe lashed to the struts. Below us is the dense primeval forest with rivers and lakes at intervals, unbroken except for an occasional trapper's cabin or perhaps a mining shaft. Sometimes we have looked down and spotted a moose feeding in a creek. It raises its head and remains still as the aeroplane approaches, hoping it may not be seen; then as the plane passes over, it scrambles to the bank and lopes off into the darkness of the bush.

Often, I have been in an aeroplane that has brought to an isolated camp the first mail and news of the outside world for months. Eagerly, two or three men wade into the water to guide the floats to the bank. The whole company stares at the rare bird and at the white woman. The cook is all hospitality and smiles. He apologizes for the meagerness of the meal, which is in reality abundant, and for the makeshift arrangements, which are rough but clean. The men devour their mail. We are given letters to post, messages to deliver, and then away again.

One trip took me to an Indian encampment on a beautiful birch-studded lake. We had gone in with a note from the provincial police directing that a papoose be sent out to receive a farewell from its mother, a woman who was dying in hospital in a town on the transcontinental line. I held the babe on my knee on the return trip and quite fell in love with it, but could not cuddle it, strapped as it was in its stiff-back cradle. Fur traders often finance trips for the Indians in the summertime and receive payment in skins the following winter. The pilot sees that all freight for air traffic is weighed to prevent exceeding the maximum load for which the aeroplane is licensed. The women seldom tip the beam at less than 200 pounds. My husband swears it.

One beautiful summer day, we received word that a prospector had died close to a camp on the Indian reserve and that provision must be made for his burial. We flew there with the coroner and a priest. A simple funeral service was held on the veranda of the log shack, the prospector's chum and a few Indians kneeling in attendance. The man who had relinquished his earthly "claims" was laid to rest beneath a tree at the lakeside. The mechanic thought it was a pity that he was buried in unconsecrated ground. I thought it was a beautiful spot and felt that the old prospector would prefer to rest in the open where he had spent his life rather than lie in some stuffy town cemetery.

One flight my husband made left me in a flat spin of curiosity, for he was not free to disclose the nature of it at the time. Two ladies had arrived from Montreal and, heading straight for the local air base, requested that an aeroplane be dispatched to a camp some 40 miles distant to bring a certain man to town.

"Who is going to pay for the trip?" asked the pilot.

"He will," replied the shorter of the two, a wiry little lady. "He'll pay for it; he's my husband."

"And he is my husband too," joined in the other, quite a buxom lady. "You bet he'll pay for it."

It seemed that the man had married them both at different times and had put his letters to them in the wrong envelopes. The aeroplane flew out to the camp but returned without a passenger. As it neared the dock, the pilot could see two heads peering around the corner of the office. He had much explaining to do to convince the ladies that their mutual husband was not in camp.

Grub, canoes, mining machinery, gold bricks, husky dogs, furs, anything and everything comprises freight loads in the north country. Once it was necessary to bring a dead Indian back in from his cabin. My husband was flying a trader on a tour of fur buying.

"We'll just drop down and see if Lazy John has anything since we are over his place," the fur buyer said. Gliding down onto the frozen lake and taxiing the aeroplane over to the cabin beside the lakeshore took but a few moments. They entered the shack and found the man

A picture of Lucille Burton, who died in 1996. *E. Burton*

dead in one corner, his wife sick and hungry, and two half-starved infants without food or fire. A meal was hastily prepared from the rations that are always carried in the aeroplane and ravenously devoured by the woman and youngsters. They were flown to hospital and the dead husband brought out later for burial.

I frequently encounter the question "How does it feel to be a pilot's wife?" but I do not believe there can be a complete answer to it. The emotions are too conflicting. One woman is envied as she drives her handsome husband to a city airport in an expensive car, smiling at the deferential attention they both receive. But who knows whether behind her smile she is rigid with fear as the pilot takes off for his daily stint of flying? Or another flier's wife is the object of pity, perhaps stuck in the middle of nowhere with her husband doing a pioneering job. But it may be that her anxiety is not greater and her recompense just as satisfying as the other's. Whatever their reactions, I know of few wives who will ask their husbands to give up the career that is their lifeblood.

Fly Anywhere*
Air Engineers Share
the Hazards of Flying

H. M. (BERT) PHILLIPS

In the years prior to World War II, the contributions of air engineers were as important to the success of any bush flying operation as were those of the pilots, who enjoyed the lion's share of public recognition. In the first pioneering decade of northern flying, an engineer accompanied every pilot, each trusting and depending upon the other. Forced landings due to equipment failure were not uncommon, and, with no radio to summon help, an air engineer's skills were vital.

When Bert Phillips arrived on the scene in the mid-1930s, aircraft engines had become more reliable. Yet engineers regularly flew with the pilots, especially on longer trips that took them away from base for extended periods. Engines required regular servicing, and damaged aircraft had to be repaired. The importance of the air engineer remained undiminished.

The truth of this statement is borne out by the following account, which deals with much more than personal experiences. The author paints a broad picture of this later, but still pioneering, phase of aviation. He tells of life in the wilderness, of cargoes carried, of equipment and its shortcomings, and of the ingenuity needed to keep the system working in times of financial stress. While Bert's employer, General Airways, operated mostly in Quebec, the circumstances he describes prevailed across the Canadian north.

---

*"FLY ANYWHERE" was General Airways' motto and a part of their logo.

Born in Cobourg, Ontario, Bert moved with his family to Toronto shortly after World War I. He grew up there and begins his story when he was 19 years of age, during the depths of the Depression. Work was scarce, and to a young man already sold on flying, the offer of a job in aviation would have seemed a dream come true.

After his bush flying years, Bert Phillips served with distinction in the RCAF, first as an aeroengine mechanic and then remustering to air crew as a flight engineer on Consolidated Catalinas in the Far East with 413 Squadron flying out of Ceylon (now Sri Lanka). Upon his release from the air force, he joined Field Aviation of Toronto, where, because of his expertise on large aircraft engines, he was seconded to Kenting Aerial Surveys. He would maintain their Cansos and Flying Fortresses, not only in the Canadian north, but also in parts of the world as remote as northern Greenland and as distant as India. His story is based upon a talk given to the Toronto Chapter of the Canadian Aviation Historical Society in June 1985.

M y entry into the world of commercial aviation occurred in 1934 at a time when I was "piloting" a bicycle, delivering for a drugstore in the Rosedale district of Toronto. I spent most of my free hours at the Toronto Flying Club (TFC), helping out and taking flying lessons. Shortages of time and money stretched my dual instruction with Chief Flying Instructor O'Brien Saint from June until August, and it was a further six months until "O'B" signed my application for a pilot's licence. His death in the crash of Gipsy Moth CF-CBK was a sad setback, not just for me but for aviation in general.

That same day I had a delivery to a famous customer, World War I flying ace Captain Roy Brown, Baron von Richthofen's nemesis. Brown answered the door, and we fell into a discussion that came around to O'B's accident. When he inquired about my personal plans in aviation, I explained my situation, and he told me that his company, General Airways Limited, was expanding and would soon have some openings. He promised to get in touch with me.

A few weeks later Roy Brown proved as good as his word. He phoned and invited me to his home to meet Wilson ("Clarkie") Clarke, general manager and chief pilot of General Airways. The outcome of this meeting was a job for me as a trainee helper at $75 per

Pilot Bert Phillips (right) with Bing Davis at the Toronto Flying Club in 1934 in front of one of the club's D. H. 60 Gipsy Moths, (G-CAJU), the *Sir Charles Wakefield. H. M. Phillips*

month. Since company policy at this time required pilots to hold an air engineer's licence, I was to continue training at the TFC under the new instructor, J. R. K. (Ken) Main, until General's newest aircraft, a Bellanca Senior Pacemaker, arrived.

It was early summer 1935 before I finally received my orders to travel to the Rockcliffe RCAF seaplane base on the river near Ottawa to meet the aircraft. Roy Brown was also there when "Clarkie" flew in with our new Bellanca, CF-ANX, on floats. A number of RCAF officers were taken up on demonstration flights—'ANX was a jump up on the older Bellanca CH-300 Pacemakers that the air force was then operating.

Eventually "Clarkie," Roy Brown, and I boarded our machine and took off for Noranda, Quebec, General's main base. Clarke was in the left seat, Brown in the right, and yours truly in the back with the baggage and such other gear as undercarriage, wheels, and fairings. There were no skis; we would obtain these later from Elliott Brothers in Winnipeg.

About an hour after takeoff, Brown called me forward and asked if I knew where we were. My guess wasn't close enough to suit him, and he lectured me on the importance of always keeping a map handy and continually pinpointing one's location, regardless of who was actually doing the flying. At the same time he stressed the wisdom of being suitably dressed—in case one had to "walk home"—no matter what the season. Thus began my training in bush flying.

All General Airways' aircraft were well equipped with survival gear: sufficient emergency rations in sealed cans to feed a full load of passengers for about 10 days, guns, ammunition, fishing tackle, cooking utensils, matches sealed in wax in a metal container, hatchet, axe, sleeping bags (for crew), tool kit, and rope. In winter we also carried a couple of pairs of snowshoes, an engine cover, and a large blow pot (for engine warm-ups).

In spite of the fact that 'ANX was the new flagship of our fleet, and that the top brass were on board, our arrival at Noranda caused no excitement. Most of the other aircraft and crews were away on charters. But I was warmly greeted. From the beginning, everyone at General made me feel welcome, and they were always ready to help a greenhorn like me whenever they could.

NORANDA BASE

One of my first and most lasting impressions of Noranda was of the twin smokestacks, some 599 feet high, at the Noranda smelter. From these belched continuous clouds of dense white smoke strongly smelling of sulphur. This odour could be detected even in the cabin of an aircraft flying many miles downwind. On a hot, windless summer day, the whole area would be blanketed with acrid, sulphurous smog, making breathing uncomfortable. Tons of molten slag were dumped daily on the lakeshore northeast of Noranda, creating irritants in the water that affected swimmers and even corroded the aluminum floats of our aircraft. The glow from the slag, however, made an excellent flare path at night and in poor weather!

General Airways' main base was on the west side of Lake Osisko, between the towns of Rouyn and Noranda. The buildings were of frame construction, painted International Orange. They housed our office, radio room, a passenger/cargo room, and a well-equipped workshop. At this time radios were land-based only; none of the aircraft were fitted with them. Other stations were located at Opemiska and on the Oskelaneo River. A large spotlight mounted on the building facing the docking area was extremely useful for night maintenance of aircraft and in the loading and unloading of cargo after dark. Occasionally I made use of the lighting for a night photo.

The dock decking rested on plank frames supported by empty oil drums, all secured in place by lengths of two-inch water pipe driven vertically into the soft clay lake bottom. Ramps gave access to these floating docks. The system had enough flexibility to compensate for water level fluctuations over the summer.

Most of our major maintenance was done at Noranda, where the aircraft were beached during freeze-up (fall) or break-up (spring) so that the appropriate undercarriage could be installed for the coming season. Complete overhauls of Wright engines were carried out under the supervision of Bill Turner, a "D" licence engineer who had had factory experience at the Wright company plant. Usually the aircraft requiring an engine overhaul would be moved up to the workshop and then the plane's nose was inserted through the wide doors at the end of the building. Gaps around the cowling were closed with canvas curtains. In the warmth of the shop we removed the engine and stripped it, piece by piece, in the reconditioning process.

This enabled apprentices, Norm Brown and myself, to acquire valuable experience in carbon scraping, valve grinding and lapping, cylinder honing, and also part-washing with gasoline. The finer work, requiring the master's touch, was done by Bill. We dismantled the engines completely. When the procedures listed were completed, we would reassemble the hodge-podge of pieces into the semblance of an aircraft engine. Once its engine was reinstalled on the mounts, we moved the aircraft to a clear area for breaking-in runs. We began with short periods at idle speed, gradually working up to full throttle. The whole procedure required four to five hours, and usually took place late at night when it always seemed to be very cold. Understandably, we apprentices won the job of sitting there half-frozen with the chief engineer doing spot visits to monitor the results.

During freeze-up we prepared the engines for the cold weather operations ahead. We removed oil coolers (to be reinstalled in the spring), "lagged" oil lines—by wrapping them with asbestos lamp-wick, then applying a coating of shellac to prevent oil soaking. The oil tanks were then covered with heavy felt, cut and stitched to fit, and were covered again with asbestos cloth that also received a coat of shellac. Cockpit heat came from a sleeve that fitted around the exhaust and fed into the cockpit through a flexible metal tube that was checked carefully for leaks. We wanted no carbon monoxide seeping into the aircraft. In the spring, during break-up, we reversed the whole procedure, stripping the insulation from the oil system.

During the spring of 1937, Norm Brown and I wrote our "A" and "C" air engineer's exams under the supervision of a Department of Transport inspector during one of their frequent visits. We both passed.

The work we did in the shops was not exclusively on engines. Ski repairs were commonplace. Our welder devised a new type of pedestal, which we fitted to our aircraft skis. The original pedestals, made of steel tubing, tended to accumulate slush, which would freeze and add weight to the ski. The new pedestals were made of sheet steel and were streamlined, looking somewhat like inverted coal scuttles. But they were very strong.

Airframe repairs were limited to fabric patching, cable replacement, and occasionally fuselage tube welding. Most fabric repairs were to fuselage bottoms, usually after crusted ice or hard snow during ski operations had cut them open. We carried out winter repairs under tarpaulin skirts hung around the aircraft to enclose an area that could be heated with blow pots. We would do the cleaning and stitching, then remove the blow pots before applying dope to the patches.

We also worked on floats—patching them or replacing complete panels as well as parts of the internal structure. Float repairs usually took place in the shop in winter, unless circumstances such as an accident dictated otherwise. Minor damage could be repaired with a small temporary patch that might, with luck, last the season.

Such a situation arose following a mishap to CF-AZH, a Stinson SR-7 Reliant, one summer night away from base. Upon takeoff, the pilot felt a bump, so he immediately decided to investigate. He landed and taxied over to a dock. There he obtained a boat and went back out on the lake to see what he had hit. He found a small boat containing several frightened people, minus its outboard motor. Apparently, it had

General Airways' base at Rouyn/Noranda following an early snowfall with four of the Bellanca aircraft awaiting changeover to floats. The mine's infamous twin stacks are in evidence. *H. M. Phillips*

been knocked off by one of the Stinson's floats. The float in question had a gash at the waterline extending through two watertight compartments. The next day I flew over from Noranda and applied a temporary patch of fabric and dope, which enabled the pilot to take off and fly back to Noranda for permanent repairs.

THE NATURE OF BUSH FLYING

Bush flying was then, and probably still is, a day-to-day business. The only time aircraft stayed on the ground was under severe weather conditions or during freeze-up and break-up. For the mechanic these periods meant taking a deep breath and diving into work that had been deferred until time was available. Water operations did not cease until the ice was too thick to be safely broken by the relatively thin-skinned floats. Since shore ice formed first, it was still possible to take off in the open water farther out. Then the pilot had to ice-break his way into shore at his destination. I have been aboard aircraft when there was no open water to land on, and the pilot was obliged to set his float-equipped machine down on the fresh ice. The ice would support the aircraft for a short distance until most of our momentum was lost. Then the floats broke through, slowing the aircraft to a stop as if it had brakes. The float bumpers would be badly scarred or even ruined to the point where they required replacement—later, after freeze-up. At that time the floats were replaced with skis, which were installed with new shock-cords. The skis would have been overhauled during the previous summer's slack period when we had time to replace worn fittings, repair damaged brass runner plates, and check cables.

As the weather warmed, the shore ice, which had been the first to form, was also the first to melt. The ice remained solid farther out in the lake for some time after the snow had departed, and pilots used it for ski landings as long as possible. By the time flying was called off, there would be a gap of from 40 to 50 feet between the shore and the ice. Then wind direction became a factor, either pushing the ice farther out, or bringing it in. The result was hard work for us and a possible soaking when aircraft had to be taxied, towed, or manhandled over the gap.

It was always a pleasure to begin summer operations—in spite of the duckings that invariably occurred when someone forgot that it was no longer possible to jump from the aircraft to the ground below. Clouds of blackflies, often so thick that they obscured your vision,

An end-of-season hazard: Pacemaker CF-ATN has broken through the thin ice near the shore. (Note that the tail-ski still rests on the ice.) The aircraft will be raised on the A-frame and worked ashore. *H. M. Phillips.*

began to appear—usually on the spot where we had to unload the aeroplane. Mosquitoes were there in full force as well, competing for their pound of flesh. Impromptu overnight stops, however, were more comfortable, and spruce boughs made good mattresses for our eiderdowns, even on rocky shores.

GENERAL AIRWAYS' STAFF AND AIRCRAFT

General Airways' flying personnel in the summer of 1935 included Wilson Clarke, chief pilot and general manager, flying the Senior Pacemaker CF-ANX; Stuart Hill, pilot at the Amos Base, flying Stinson SR-5 Reliant, CF-AWO; Kelly Edmison, flying Bellanca CH-300, CF-AOL; Gath Edward, flying CH-300 CF-ATN; "Bun" Paget, flying CH-300, CF-AEC; Tim McCoy, pilot at our Hudson, Ontario, base, flying Pacemaker CF-AND; and Curt Bogart, pilot of our D.H. Fox Moth, CF-API.

Our engineers were Ross Baker, with a "B" licence, based at Amos; Bill Turner, "D" licence, chief engineer at Noranda; and Aime Stalport, "B" licence, Noranda. The rest of the engineers were based as required and not necessarily attached to any particular aircraft. An engineer usually went along with the Bellancas to assist the pilot with the handling of cargo and to service the aircraft at its destination. These men were George Millham, Tony Bruneau, Norm Brown, and myself. Jack Jones, based at Noranda, was a specialist on sheet metal work and welding.

Our radio staff included Pete Casey, Bert Crump, and George Pope, while the base manager at Rouyn/Noranda was Howard Brown. Later arrivals were Pat Twist, Tom Mahon, George Ward, and Dick Gunter, all pilots, and "Red" Sarsfield, a mechanic.

Our aircraft were painted International Orange or Chinese Red with the usual large black registration letters. This colour scheme was chosen because of the high visibility it afforded, summer and winter. An exception was our Norseman CF-BAN, delivered later, which was silver with a red speed line and wide red bands chordwise on the outer wings.

Bellanca aircraft constituted the bulk of our fleet. Their starting procedure required a five-armed strong man! The inertia starter, energized with a hand crank from the pilot's seat in the cockpit, took a couple of minutes to build up starting speed. Once this was achieved, the crank was removed and a foot pressed against the shaft to engage the engine. As soon as it turned over, the switches were flipped to *on*. The throttle was set and a free hand used to crank the booster magneto. If all was well, the engine started. If not, the frantic process had to be repeated in its entirety. Prior to all this, the propeller had to be pulled through several times to clear and prime the cylinders. Occasionally you could start the aircraft by hand-swinging the propeller, especially if the plane was on skis and the engine warm.

Our de Havilland D.H. 83 Fox Moth, with no starter, always had to be hand-started by swinging the prop. An auxiliary set of switches and throttle was mounted on the left side of the engine cowl for easier access by the prop-swinger, if he were alone.

FLYING ANYWHERE, DELIVERING ANYTHING

From Rouyn/Noranda we flew "anywhere" with regular runs to Val d'Or and the Chibougamau/Opemiska regions of Quebec; from Hudson, Ontario, we serviced the Red Lake region; from Amos, Quebec, we did the Siscoe/Malartic area; and from Oskelaneo River, Chibougamau. The base at Amos used the Harricanaw River for summer float operations and a farmer's field east of the town for winter flying on skis. This field had the only hangar that could accommodate a Stinson Reliant, but not the larger wing span of a Bellanca.

Canadian Airways Limited, situated on the north shore of the lake, provided our competition at Noranda. It operated the one-of-a-kind Fairchild Super 71, CF-AUJ (usually flown by "Red" Lymburner); the Fox Moth CF-APG, and usually a Fairchild 71. A closer neighbour was Johnnie Fauquier, who operated a very rare Fairchild 51/71, CF-AWP and CF-AUR, a Waco UKC (with a Continental engine). Dominion Skyways was located on the southwest shore at Rouyn and had the prototype Norseman Mk I (450 hp Wright Whirlwind) CF-AYO; a Fairchild 82, CF-AXB; a Bellanca CH-300, CF-ANI, and a Fairchild FC-2W-2, CF-AUG. The rivalry that existed between General and the above firms was limited strictly to commercial activities. We assisted one another freely, and friendships abounded among the personnel of the various companies. It was a pleasure to be part of the brotherhood that existed among those involved with bush flying.

The most common flying routes were Noranda to the Chibougamau area, about 250 miles, or two-and-a-half to three hours

General Airways' logo with their once well-known motto "Fly Anywhere." The colours were black, red, and white. *H. M. Phillips*

by Bellanca cruising at 90 miles per hour; Noranda to Val d'Or, about 70 miles or 50 minutes; Amos to Val d'Or, 20 minutes; Noranda to the Oskelaneo River, about 200 miles or two-and-a-bit hours; Oskelaneo River to Chibougamau, 125 miles and one-and-a-half hours. General Airways' motto was "Fly Anywhere," and to this could be added "Fly Anything," provided it could be loaded into the cabin or secured to the floats. While we did not have scheduled flights, trips to certain areas were frequent. General serviced a mail contract from Rouyn to Malartic with as much regularity as the weather would permit.

Some of our customers used the company address and would receive their mail with their next re-supply of food or equipment, or we would deliver it to them if we happened to have an aircraft flying nearby. This policy was also applied to prospectors who had been dropped in an area with a pick-up date set for weeks or months later. Items for them would be left at their campsite. We also monitored these people in case an emergency had arisen and they needed the aircraft. We made low passes over their areas, looking for a friendly wave-off or a more frantic signal to land. This sort of communication was important because their locations were often known only to us, kept secret from their competitors for obvious reasons.

SOME EMBARRASSING LANDINGS

In summer almost any water could serve as a landing place, even swampy ponds. This sometimes resulted in interesting takeoffs, involving racing around in large circles on one float to build up lift-off speed. This could also happen in winter when the same surface froze over. Even prepared strips might pose problems. For convenience, a mine might clear an area near its property that would be used for ski operations once there was sufficient snow. O'Brien Mines, a few miles east of Noranda, had such a strip. Our Stinson SR-5, CF-AWO, had been damaged there before I joined the company.

Opemiska Mines also had one of these strips. In December of 1935 it was not completely covered with snow, and this contributed to an accident involving Bellanca CH-300, CF-AEC, piloted by "Bun" Paget. It was the first flight after freeze-up, and, before leaving Noranda with a full load of passengers and cargo, one of the people on board had asked "Bun" how close to the camp he would land; apparently this passenger did not like tramping through the snow. "We'll drop you right at the door," "Bun" assured him.

I was on board the flight, and as we were about to set down, "Bun" realized that there was not enough snow. So he decided to abort the landing. But the engine was slow to respond and didn't pick up until we hit the tops of some jack pines in a ravine off the end of the runway. Then it came to life and pulled us through the trees right up to the higher ground at the mine where our right wing caught the radio pole and swung us into the radio shack. Like the rest of the buildings there, it was a good solid log structure and suffered little, unlike 'AEC, which was badly damaged. When the dust and snow had settled, we were pleased to find that no one had so much as a scratch. "You really do live up to your word, don't you!" commented the passenger who didn't believe in unnecessary walking.

The company decided to salvage 'AEC so it flew in George Milham, an air engineer. George and I dismantled the plane and loaded the pieces onto a freight sled. The next tractor train hauled it south to Senneterre on the winter road, and from there it travelled by rail to de Havilland's at Toronto to be rebuilt. This was my first experience of salvage—and of mining-camp life. The food was excellent and plentiful. Our bunkhouse, where the temperature ran at about -20 (F), was less impressive. The small stove died soon after we climbed into our sleeping bags. The water bucket always froze overnight, and snow blew in through the poorly chinked log walls and had to be brushed from our eiderdown bags in the morning. Toilet facilities also left a lot to be desired.

Malartic Mine was also equipped with one of these bush strips, and it was here that 'ATN hit a snow-obscured stump, damaging the undercarriage and one wing. We removed the wing and crated it into the Malartic carpenter shop for shipping to DH at Toronto. 'ATN received new skis, and when the rebuilt wing arrived it was re-installed. Gath Edward test flew it from Noranda to the Siscoe Mine, where, the evening before (7 March 1936), our Reliant, CF-AWO, had taken off with no one aboard and crashed.

Stuart Hill, pilot of the ill-fated 'AWO, had left the engine idling since he was stopping only briefly—with a passenger in the right-hand seat—while he went to the mine office to make his delivery. The passenger began to feel the cold and decided to climb out and move around to restore circulation. As he did so, his foot kicked the throttle open, and the engine roared. Frightened, he tumbled out the open door just as the empty aircraft began to move forward. It gathered speed and took off, climbing to about 300 feet. Then it made a turn

Bert Phillips with Pacemaker CF-AEC after its 1935 crash on the mine strip at Opemiska, Quebec. The aircraft was rebuilt and flown until 1942. *H. M. Phillips*

and dived back into the ice—almost at the spot where it had been parked. This was 'AWO's second accident, and this time it was a write-off.

In February 1937 Ross Baker, the base engineer at Amos, was taken to hospital with appendicitis, and the company sent me to fill in for him. As I have noted, Stuart Hill was the pilot at Amos with the Reliant 'AWO. Since it was winter, we were operating from the previously mentioned farmer's field outside of town with the small, unheated hangar. Attached to the hangar was a lean-to structure containing a workshop and a tiny office with a stove on which there was always a large pot of coffee—a great substitute for varnish remover! Life here moved at a less hectic pace than at Noranda, although the Reliant made frequent flights to the mines at Malartic, Siscoe, and Sullivan.

FIRE! ALWAYS A HAZARD

One afternoon, four of the Noranda aircraft arrived at Amos to wait out a snowstorm that had closed down their base. Although the pilots pitched in, tying their individual aircraft down, this substantially increased my work. I had to drain the oil from each aircraft and take it into the warm workshop, where it would be reheated before being poured back into each engine once the weather cleared. Engine and wing covers also had to be put in place on each machine. When, after two days, the storm finally cleared, the pilots were anxious to get airborne at first light. With the passing of the front, the temperature had dropped from merely cold to very cold. And so I was out at the field before dawn, getting the stove going, setting the cans of oil on it to warm, and taking off the wing covers and packing them into each aircraft. Then I proceeded to place blowtorches under each engine-cover. There were no purpose-designed Herman Nelson engine-warmers in those days. Normally the engineer would sit under one aircraft at a time inside the tent, occasionally pulling on the prop, since free movement would indicate the engine was sufficiently warm to pour in the oil and fire it up. But I was being an eager beaver—sitting outside, trying to watch them all at once.

Suddenly the oil-soaked engine cover of 'AEC burst into flames. I grabbed the Pyrene fire extinguisher, tore the cover off the aircraft, kicked the blowtorch out of the way, and managed to put out the fire.

Bert Phillips with General Airways' Bellanca Pacemaker, CF-AEC, at Noranda, Quebec, in 1935. *H. M. Phillips*

But the engine cover was ruined and so was the ignition harness on the Bellanca's engine. Fortunately, the other three aircraft were able to depart at first light.

Now there was a set of ignition wires to be replaced on an aircraft that was too big to fit into the hangar. We pushed it up to the doors with the nose inside as far as it would go. Then we closed the doors against the cowling and covered the remaining open space with tarpaulins. This allowed us to work inside, out of the weather.

Unfortunately, this left our Reliant outside overnight, and it had to be heated again in the morning. This time I was able to sit under the cover with the blowtorch. All at once the torch spurted a jet of burning gas up and onto the cowling. Hurriedly closing the valve on the torch, I put out the fire, but not before the dope had burned from the cowling. The company was going to fire me until Ross Baker heard the story and spoke up for me. It seemed that he had modified this particular torch so that it could burn avgas instead of the usual naphtha. When I used naphtha in Baker's modified torch it did not

77

produce enough heat to burn all the fuel, and some of it sprayed out raw—and then ignited!

Fire seems to have been a recurrent problem at Amos. A couple of years before, the company lost its Fairchild FC-2 Razorback, G-CAJJ, when it too caught fire during the morning warm-up. This was in February 1935. Previously, their Fokker Super Universal, CF-AEW, had been destroyed by fire in July 1930 when flames from the exhaust had ignited the fabric during a takeoff from the river.

AN EXTENDED CHARTER

In June of 1937, a group of geologists chartered our Norseman for an exploration trip to Richmond Gulf on Hudson Bay. Kelly Edmison was the pilot, and I accompanied him as engineer. The flight north to the gulf was about 650 miles. We had a radio (HF), which did not perform well. But once we landed at Great Whale River, Kelly was able to get word to Noranda that we were okay. During the trip we refuelled from RCAF gas caches through a prior arrangement with the Department of National Defence. The Norseman burned about 30 gallons an hour at a cruising speed of about 110 mph. We refuelled at Fort George at the mouth of the Grand River on James Bay, and continued north along the coast.

About 75 miles north of Fort George, at the point where Hudson Bay begins, a line squall lay squarely across our track, stretching to the east as far as we could see. Since it was moving toward us, we did a 180 (turn) and headed back to Fort George, just making the beach as the storm hit. This meant an overnight stop—until the storm passed. The Indians there staged a dance in our honour. The dance took place in one of the Hudson's Bay Company storage sheds and lasted most of the night. The band consisted of a concertina, mouth organ, fiddle, and a small but noisy drum. It was not a large room, and, with 40 people in it, the drum and dancers made enough noise to drown out the thunder outside.

The next morning we were on our way again. Fog almost obliterated the coast, and we had to fly the next leg to Great Whale River at about 50 feet above the rocks and waves. Thanks to Kelly's able handling of the Norseman, the flight was uneventful. After we landed on the river at Great Whale we drove the Norseman as far as possible onto the sandy beach and tied up to some bushes.

The Hudson's Bay staff and the local Eskimos received us warmly, fortunately, because the dense fog grounded us there for several days. Finally, the wind reversed and blew the fog out to sea—less fortunately it also kept the tide out, leaving us high and dry on the beach. After laboriously turning the aircraft, we enlisted all of the local Natives to haul on ropes attached to the floats. Then, with the engine roaring at full bore, we were able to reach the water and float the machine. The flight to Richmond Gulf, less than an hour to the north, was a pleasant anti-climax. On the gulf we camped by a trout stream, and a feast of freshly caught game fish, fried with bacon and supplemented by the usual wieners and beans, left us well fed and content.

The weather remained fair, and our passengers left on their field trips. Although they found some commercial diamonds, these were not in sufficient quantities to make mining them a paying proposition. Once they were satisfied that their search had been completed, we packed up and headed back to Great Whale.

When we returned to Fort George we found that, in our absence, a four-year-old Eskimo girl had been mauled by huskies and needed medical attention. Quickly refuelling, we loaded the youngster and her mother on board for the flight to the hospital at Cochrane, where the landing area was a shallow lily pond, almost a swamp. As we set down, our floats left tracks in the weed-choked water. But we delivered our charge safely. And, happily, there was no problem getting airborne. About an hour later we were back at Noranda after only a month's absence. In 25 hours flying time we had experienced no snags with either the engine or the aircraft.

CARGOES, OFTEN CHALLENGING, SOMETIMES BIZARRE

The loading of an aircraft was always at the discretion and direction of the pilot flying the trip. Objects such as lumber or pipe, if they were too long to fit into the cabin, would be lashed to the float struts—as would a canoe. The drag of one canoe was equated with the weight of half of a payload, and the cabin load was decreased accordingly. But loads often exceeded what was permitted: the best indicator of actual weight was the water line on the floats. If they were almost awash, you were in trouble. Pilots logged true weights until inspectors from the newly formed Department of Transport began checking the logbooks and handing out licence suspensions. Then the figures in the logs began conforming religiously with those posted for the particular air-craft types. Some materials were compact and heavy; a normal load of diamond drill rods did not take much space—and neither did an overload.

Dynamite was another item that lent itself to overloading. It too was compact, and cases fitted neatly in the cabin. While fresh dyna-mite posed little risk, if it had been frozen or was very old, the nitroglycerin could separate from the sticks, forming little beads of moisture, and raw nitro *was* dangerous. Usually it was part of the gear being moved from one camp to another. If it showed any sign of leak-ing, we would not touch it. Yet one summer we flew two entire boxcar loads of explosives from our Oskelaneo River base into Chibougamau without incident.

Our loads varied: we might have a trapper, his food, camp gear, dogs and sled—or we might have a local hockey team complete with coach and equipment. Nor was it uncommon for a couple of enter-prising young women from Rouyn to charter an aircraft for a flight into a busy mining, lumber, or road camp. Their baggage would con-sist of many cases of beer and hard liquor, tents, cots, and a cash box—no Visa cards! Such visits usually coincided with paydays.

Some of the Amos flights entailed picking up gold bricks or "saucers" depending on mine output. The mines coordinated the pickups and alternated in providing security guards. The first mine visited would supply such a guard, and he would remain with the shipment as they were successively collected on the way to the railway express office. The gold was packed in sealed boxes, and at Amos these were unloaded into a snowbank to await the arrival of a taxi from town. There might be as much as $100,000 worth of bullion sitting outside while everyone was inside enjoying coffee. The taxis were ancestors of today's snowmobiles: the front wheels had skis attached while the rear wheels were paired in tandem with a belt fitted over each set like the tracks on a caterpillar tractor. Otherwise the vehicles were regular sedans.

The Quebec government was another charter customer in early winter for trips to the prison at Ville Marie. The passengers they spon-sored were girls from Rouyn who had been picked up during police raids. There would be about 30 or 40 of them who could not, or would not, pay their fines, and so ended up on the shores of Lake Osisko in groups waiting to be taken south. The rest of them waited in our passenger room, unescorted. They seemed to regard it as great fun and were probably looking forward to the rest.

Sometimes the bodies we carried were not so warm; one in particular belonged to a man who had died of exposure on the trail back to the mine where he worked—after a visit to a house of ill repute in Val d'Or. By the time he was discovered, in temperatures well below zero, his body had frozen solid. Since we had passengers for the flight he was wrapped in canvas and became a seat. No one wanted to wait for the next flight, which might be held up for days by bad weather.

On another occasion we had a rush call to pick up an injured miner from the same area. This poor fellow had been drilling underground when his bit ran into an unexploded charge. The resulting explosion drove a four-foot piece of drill through his shoulder and riddled his flesh with small rock chips. When we arrived he was conscious, though sedated. We wasted no time in getting him back to the hospital at Noranda where the steel was removed and the rock particles scrubbed out of his skin. He survived.

While some loads might seem typical and easy to handle, their disposition could be unusual. A large mining company had us move tons of rock samples from a tiny lake about 30 miles from Noranda. The rock was in canvas bags of a size easily handled by one man, and altogether they made quite a large pile. Strangely, only about a half-dozen of these bags were ever collected by a representative of the company. A couple of weeks later we received instructions to empty the rest of the bags and dispose of their contents as we wished. It became ballast for our ramps—*all that glitters is not gold.*

One summer an unfortunate government surveyor drowned near Opemiska, and his body was not found for several weeks. The corpse was then wrapped in a tarp and left on the beach for an additional week before we were requested to transport it to the coroner at Amos. Fortunately there were no passengers on this flight, and we flew with all the side windows open and our noses in the slipstream—for obvious reasons. Some weeks later we began receiving inquiries from Ottawa regarding a missing government tarpaulin. We in turn prodded Amos and then forgot about it. At about the same time we had made a request to Noranda for engine and wing covers for our aircraft. When a large bundle arrived wrapped in heavy brown paper I thought of our new covers—but a small tear in the paper emitted an odour that told my nose what had once been inside. The bundle was left intact for the pilot's inspection and, unfortunately for him, his nose was not as sensitive or he was more eager to get at those new covers. He ripped it open, and the smell that assailed him made him violently ill. We did not forward the tarp to Ottawa, although the idea was tempting. We burned it there on the beach with the help of a generous soaking in avgas.

I have mentioned diamond drill rods as heavy cargo. They sometimes posed another problem. Steel in such quantity could convince the compass needle that north was somewhere back in the cabin, regardless of the actual heading of the aircraft. Other large metallic cargoes—compressors or generators—had a similar effect and would render the compass temporarily unreliable. It was at such times that map reading became most critical. And maps in those days invariably had large areas marked "UNSURVEYED, INFORMATION NOT RELIABLE." I am sure that most of our pilots, however, were so familiar with the country that they would notice if a tree had recently been felled.

A TYPICAL AIR ENGINEER'S ASSIGNMENT

About this time word came from Noranda advising me that it would no longer be necessary for me to fly with the aircraft. As I stood on the dock watching the Bellanca Pacemaker 'ATN take off, I heard a whistling sound, indicative of a burned valve. Blissfully unaware, the pilot continued on his 125-mile flight north, a good part of it a "dry hop" over a large area almost devoid of lakes and thus with few places for a safe emergency landing.

"Red" Lymburner, later to earn fame for his Antarctic exploits, piloted Canadian Airways' Fairchild Super 71, CF-AUJ, and had agreed to carry me with my spares and tools in his machine, which had ample available space. The Super 71 was one of a kind. The pilot sat in an enclosed cockpit above and behind the wing, which sat just on top of the tubular all-metal fuselage. The passengers and cargo were in a cabin below the wing and ahead of the pilot. The arrangement was similar to our much smaller biplane Fox Moth, although the Fox afforded the pilot better visibility.

While waiting for me, the pilot of 'ATN had parked the aircraft at the dock of an establishment run by a lady familiarly known as "Box

Car Annie" whose business was selling beer and liquor and afterwards providing a relaxing sauna—followed by a numbing plunge in the lake. Unfortunately her dock had not been built with floatplanes in mind; I had not only a cylinder to replace, but a punctured float to repair, the result of an encounter with a sharp-ended log just under the water. Both jobs were easily completed, and I was fortified with a good meal, a sauna and a sound night's sleep before the trip back to Oskelaneo on the following day.

Many of the places where loads were to be delivered had no dock, and float planes could not always get close enough to shore because of boulders or trees at the water's edge. This was where an axe and bucksaw came in handy. Trees could be felled to clear the area and remove obstructions for the wings. Then their trunks could be used to build a temporary docking facility. For heavy items—compressors or 45-gallon drums—stripped tree trunks were lashed together to form a ramp up to the floor of the aircraft. The cargo was then wrestled ashore, invariably by the engineer, who always got soaked.

THE BEGINNING OF THE END

In May 1936, General Airways suffered a disaster. Bellanca CH-300 CF-AOL crashed south of Opemiska killing "Clarkie," the pilot, George Millham (his mechanic), and five passengers. "Clarkie" had been giving George his final check ride prior to George's graduation as a full-time pilot. The opinion of experts who visited the crash site was that an unusually severe downdraft had been the probable cause. Although we did not realize it at the time, "Clarkie's" death marked the beginning of a gradual decline for General Airways. Within months our aircraft were being sold off.

The loss of these pilots and their machine brought about changes in aircraft allocation. Gath Edwards had flown CF-ATN and Kelly Edmison CF-AOL. Now they shared the Senior Pacemaker CF-ANX. Later, when our Norseman CF-BAN was delivered, Kelly flew it and Gath had 'ANX exclusively. Curt Bogart went from the Fox Moth CF-API to the gull-wing SR-5 Reliant CF-BAG, and Stuart Hill took over the gull-wing SR-7 Reliant CF-AZH at Amos. "Bun" Paget had the newly rebuilt Bellanca CF-AEC, although from time to time, he would fly one of the later Stinsons. The newer pilots fitted in wherever they were needed.

Wilson ("Clarkie") Clarke spins the prop of a visiting Bellanca. Wearing the long coat is Roy Brown, owner of General Airways and famous as Baron von Richthofen's World War I nemesis. *H. M. Phillips*

During the summer of 1936, our de Havilland D.H. 83 Fox Moth CF-API was leased to Mike Thorne, an American pilot/prospector who used it to transport his men to and from the various mineral sites in which he had an interest. Late one evening, as he taxied to the beach at one of his camps, a man jumped onto a float and was badly injured by the propeller. Thorne immediately placed him in the cabin and took off for Amos, where he attempted a landing after dark on the Harricanaw River. He struck a deadhead and overturned. Thorne was able to get both himself and his injured passenger safely ashore and then to hospital. De Havilland repaired 'API and sold it to an operator in the west, Ginger Coote Airways.

General Airways continued to dispose of its equipment. In 1937 we sold our Stinson SR-7B, CF-AYW. The same Mike Thorne was flying it when he ran out of fuel near Noranda and made a forced landing in second growth bush near the town. The aircraft was badly damaged and went south to the U.S., where presumably it was rebuilt since it received an American registration.

In May 1937, CF-AZH, another of General's SR-7 Reliants, was sold to the Turnbull Fishing Company of Flin Flon, Manitoba. Archie Turnbull was owner/pilot and engineer of the firm. Working for him was Ken Main, nephew of J. R. K. Main. They operated the aircraft between Flin Flon and the north end of Reindeer Lake, the site of their trout fishing business. In December 1937, returning to Flin Flon, they encountered a heavy snowstorm and the engine quit. They dead-sticked onto Laura Lake on the Manitoba/Saskatchewan border. Unfortunately, some trees growing along the shore caught a ski during the approach, just as the engine decided to restart. It was too late, and the aircraft struck the ice in a nose-down attitude and burst into flames. The two pilots managed to make a hasty exit. But all of their emergency equipment was destroyed in the fire, and they were obliged to spend a cold night on the lake. Fortunately Tom Lamb from The Pas spotted them the next day and picked them up.

In January 1938 Turnbull returned to Noranda and purchased another of General's aircraft, this time the Bellanca CH-300 CF-ATN, and he also hired me as engineer. I left Quebec and began the final phase of my bush flying years—but that is another story.

GENERAL'S PILOTS WOULD HAVE LOVED THIS AIRCRAFT

An old saying has it that "Given enough power, you can fly a barn door." In the summer of 1977, I had the opportunity of seeing just what power to spare could do for an aeroplane. I was riding in a Turbo Beaver, which is certainly no barn door but does have ample power. Wes Watson was flying our aircraft, CF-UBN, and we were exploring near Schneider Lake in northern Manitoba by the Northwest Territories border. Because of the length of our trip and its urgency, Wes piloted one leg and I flew the next. We were on wheeled floats, and I could not help comparing this aircraft with the Bellancas that I had known so well. For the first time I was in an aircraft that although comparable in size to those machines had almost twice the power. It made getting on the step a piece of cake. Getting off the step into the air was even easier. With reversing thrust, landing and parking the Turbo Beaver was something to write home about. How the pilots of the '20s and '30s would have loved this aircraft!

West Coast
Fisheries Patrol

*A Prewar Flying Assignment
that was Challenging
and Enjoyable*

C. G. (GORD) BALLENTINE

Gord Ballentine, at time of writing, a hale 931/2 years of age, began his aviation career over 70 years ago when he obtained a job as a crewman with World War I ace Don MacLaren's Pacific Airways.

After four years crewing, repairing aircraft, and instructing would-be glider pilots, Gord earned his private and commercial pilot's licences in the same year, 1931, just as the Great Depression was worsening. Staying in the business during such times demanded ingenuity of the type demonstrated by Aircraft Services of BC, in which Gord was a partner. The "firm" was really a booking agency with access to three aircraft of various sizes, each owned by a different person. These machines consisted of a two-place, Fleet 2 open-cockpit biplane on floats, an Eastman E-2 Sea Rover open three-place biplane flying boat, and a Boeing B-1E—an enclosed five-place biplane flying boat. The nature of each charter dictated the aircraft to be used.

Gord did all of his flying on Canada's west coast, where the mountainous terrain and the ever-changing weather blowing in from the Pacific Ocean added to the hazards. Understandably, many British Columbia and Yukon pilots eschew the term bush pilot—they are "mountain pilots."

During his flying career Gord Ballentine became closely acquainted, whether as pilot or crewman, with a wide variety of aircraft. As well as the previously mentioned Aircraft Services

86

machines, his list includes Pacific Airways' Curtiss HS-2L, Western Canada Airways' Vickers Vedette, and Boeing B-1D flying boats. Later Gord would fly Canadian Airways' seven-place Bellanca Pacemakers and smaller Waco cabin biplanes, de Havilland Rapides, and Dragonflys, as well as one of their Junkers W. 34s. His largest charge was a Canadian Pacific Airlines' 14-passenger Lockheed 14H. Gord Ballentine may be the only pilot alive who can claim extensive working experience with so many machines that are now, for the most part, found only in museums.

Gord's final position in aviation was as personnel manager with Pacific Western Airlines, forerunner of Canadian Airlines International. From crewman to airline executive had taken three decades, less a few years away from flying.

Gord Ballentine prepared this story for the Fall 1993 Canadian Aviation Historical Society *Journal*. He writes with verve and humour and with his own unique style. His ability to see the amusing side of difficult situations that tested his judgement and taxed his skill is a trait he shared with many others who flew during that pioneering era. It probably enabled him to survive and write about it. He begins his story in 1937 shortly after his return to Canadian Airways Limited (CAL), as Western Canada Airways, WCA, was renamed, when it began to expand its British Columbia operations.

"ACCORDING TO OPERATIONS ORDER #1, PROCEEDED ON DELIVERY FLIGHT TO SWANSON BAY. LANDED AT ALERT BAY FOR FUEL..."

Typed in quintuplicate on legal-sized paper, *Air Patrol Report SB-1* records day one of a job I have often said was the best I ever had—at least with a seaplane. Must have been or I wouldn't remember it with such nostalgia. As much fun as work, like playing cops and robbers. Bill Jacquot and I were off to help police salmon fishing on British Columbia's north coast. SB-103 would bring us home some months later.

Had anyone asked, "Are we having fun yet?" he would have gotten a grumble from me. I'd been married just two months and couldn't take Betty along because our main base, Swanson Bay, was an abandoned and isolated mill and townsite, from which we'd be away for days at a time.

To get ahead of things for a moment—when the job proved to be so satisfying, I asked for it again in 1939. Then, during the winter, I convinced the Fisheries Department in Ottawa that Bella Bella should become the main base for a lot of good reasons: better weather, fuel supplies, etc., plus one I neglected to include in my brief. Betty could live there in a cottage. Bella Bella was no metropolis, but there were a few families there, including the inspector's, as well as a hospital on a nearby island. And the RCAF was to anchor a scow around the corner in Gulichuck, complete with radio, workshop, and living quarters.

Apart from these differences, the 1938 and 1939 seasons were much the same. I flew Bellanca Pacemaker CF-BFB (an ex-RCAF machine) in 1939, with Roy MacDougal as my engineer. The start of World War II in September caused some staffing changes. As a result, I had to finish off the Southern District after mine was done. Larry Dakin, my engineer, and I worked out of Alert Bay and Nanaimo, British Columbia—with a Stinson "gull-wing" Reliant CF-BEB and a Pacemaker CF-BFB.

BC FISHERIES PATROL

Back to day one, 1938—with our Pacemaker, CF-BFC (also ex-RCAF), loaded with personal gear, food, spares, and so on—Bill and I arrived at Swanson Bay, British Columbia, after a comfortable flight. Good weather to Alert Bay, then under a ceiling that kept lowering until, by the time we passed Klemtu and China Hat, it was about 300 feet. Just low cloud and drizzle, not stormy.

We had both done time in Swanson Bay in past years; so after renewing acquaintances with Ross and Roy Tapp, caretakers of the site, we took up residence in what was the last habitable house. A couple of days later, Inspector Ed Moore arrived in his vessel, the *Senepa*, to discuss operations with us. The air crew's job, in brief, was to patrol, often in cooperation with a service vessel, watching for violations of regulations during the fishing season and then, in the fall, to finish off the contract by helping inspectors with their annual count of spawning sockeye salmon.

The Air Board, which became the RCAF in 1924, was the first to provide an air patrol. When Donald R. MacLaren formed Pacific Airways in 1925, he applied for and received the contract. Beginning

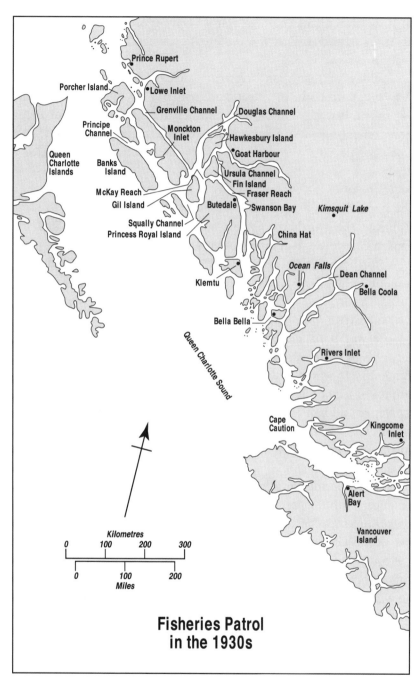

**Fisheries Patrol
in the 1930s**

The British Columbia coast in the 1930s,
fisheries patrol area. *B. McLelland*

in 1926, he, with air engineer Harold Davenport, did the patrol with an HS-2L, G-CAFH. That was the situation when I came on as a crewman in January 1928.

In May of that year, Western Canada Airways bought Pacific Airways, and both the fleet and fisheries contract were greatly expanded. It was common that summer to have three or four flying boats operating out of Swanson Bay. It was a mixed lot: the old HS-2L, Boeing B-IEs, and a Vedette. The staff there of 10 or so included a cook.

In the 1930s the contract shrank to one aircraft and crew in each of the two fisheries districts: the Southern District, Vancouver to Cape Caution; the Northern, Cape Caution to Prince Rupert, and the Queen Charlottes. It remained that way for my two seasons as pilot. The Northern District was mine. I don't know how many hundreds and hundreds of miles of sea and lakeshore that takes in, but it was all mine. Terrific!

My district was divided into six areas, each with an inspector. Ed Moore was based at Butedale, Gordon Reade at Bella Bella, Reg Edwards in Lowe Inlet, Charlie Lord had Rivers and Smith Inlets, and Engle Urseth was at Bella Coola. We never did meet the sixth inspector, based in the Charlottes.

Better add half an inspector. I had been sworn in as a deputy or something—so that I could make an arrest if patrolling with no inspector on board. Come to think of it, I have never been unsworn or defrocked.

Each inspector had a patrol boat, fairly substantial, with a skipper, engineer, and cook/deckhand. The boats were big enough to have two spare bunks for the air crew if we were overnighting. In order to share the time according to need, one inspector was responsible for giving me orders. In 1938 it was Ed Moore, his Butedale base being only 16 miles from our main base. In 1939 Gordon Reade had the job. Here is an example of how it worked, and of one day's patrol; Ed Moore had decided the Lowe Inlet inspector should have some airtime. According to Operations Order SB-14:

Left Swanson Bay and proceeded on patrol via Fraser Reach, McKay Reach, Wright Sound, Monckton Inlet, Principe Channel, to Mink Trap Bay. Landed. Reported to Inspector Edwards aboard

Bellanca Pacemaker, CF-BFC, of Canadian Airways, moored on Mooto Lake, British Columbia, while the crew and passengers return from catching their supper. Left to right: Mike Cramond (guest), Roy MacDougal (engineer), and Bill Marshall (skipper of the fisheries launch). *C. G. Ballentine*

"Audrey W." With Mr. Edwards took off and patrolled via Bare Lake and west coast of Banks Island to Bonilla Arm and return via Cotby Bay to Curtis Inlet. Landed and went aboard "Audrey W." Took off and patrolled via Principe Channel, Monckton Inlet, Fin Island, Squally Channel, Union Passage to Lowe Inlet. Landed. No violations observed.

Signed - C. G. Ballentine, Pilot

For part of the day, Bill Jacquot and I patrolled alone, and for the rest of the time we had Edwards aboard. While he was with us, his vessel did a surface patrol, so we rendezvoused the second time at a different place. What the formal report doesn't describe is the freewheeling way in which we could and did fly. High, low, here and there to look at anything that might be interesting.

FISHERIES INFRACTIONS

If you drop a line or set out a net (not a mega drift net) in the middle of the Pacific, your chance of catching salmon in worthwhile numbers is somewhere between remote and zilch. Wait until hordes of them are streaming to the creeks from whence they came, and your chances are good to too good. Hence the Fisheries Department—which hopes plenty are harvested, but insists that enough escape to spawn and perpetuate the resource.

To ensure this, fishing may be limited to so many hours or days per week in specific areas—some creek mouths are always closed to all fishing—and gear must conform to what is spelled out in the regulations. The air patrol was not concerned with the trawler or with fish traps, only with net users such as the gill-netter and the purse-seiner.

The gill net is so called because a salmon too big to go all of the way through the net can't back out—its gills are caught. A gill net must not exceed a specified length and must be a designated mesh size. It has glass or wooden floats along the top edge, weights at the bottom edge, with the boat at one end and a buoy at the other—both lit at night.

Properly, the gill net is let out buoy-end first, then reeled off the drum as the fisherman drives his boat across the channel or river until it is reasonably straight, at which point he switches off and lets the whole shebang drift with the current. This is supposed to give salmon a fair chance—they can go under, around either end, or, if small enough, right through the net. It ceases to be a fair chance when the fisherman puts his crewman ashore with the buoy, then drives his boat across the channel as far as possible. That removes one option for the salmon, and if a cooperative pal ties his net to the opposite shore, the two can close off the entire channel.

This is pretty common practice in some areas, but it's tough to get a conviction because the man ashore would use a slipknot and yank it free as soon as he heard us. Gill nets so released from tension develop an obvious wrinkle. We used to see lots of that. Except for one net, one day. We had approached Douglas Channel via a parallel channel, hopped over Hawkesbury Island and down over 14 gill nets all nicely wrinkled and one still stretched tight. As I landed, a frantic crewman was wrestling with a jammed knot with one hand and trying to pull his pants back on with the other. Lucky for the guy, Bill and I were patrolling alone, because my sense of humour got the better of my sense of duty.

Other gill net infractions were easier to handle from the air. A net too long, or set in a closed area, or when all fishing was prohibited, meant a landing and an arrest.

Regarding witnessing such an infraction from the air, here is a bit of legal trivia. In 1928 we progressed from open cockpit HS-2Ls and Vedettes to Boeing B-IEs, which closed us in and protected us from wind and rain. But when we spotted an infraction taking place down below, the inspector required the window to be wide open. Seems that long ago in England a man was passing a pub (dead broke, obviously) when he saw through a window what he thought was a murder taking place. He didn't go inside for a closer look, but was willing to testify to what he had seen. His evidence was not admitted—because in those days, glass resembled pop bottle bottoms, so distorting his view that the judge would not accept his testimony. Our inspector wasn't going to risk a 1928 judge remembering that old case.

The purse-seiner, like the gill-netter, may break the rules by fishing in a closed area or during a closed period, or by using a net that does not conform to regulations. But the purse-seiner was every fisheries officer's main concern—and thus ours—because a purse-seiner is capable of doing more damage than any other fisherman.

The seiner sets out his net with a man in a dory, then manoeuvers so that the net is in a large circle surrounding a school of fish, with the dory back at the boat. Next a running line at the bottom edge of the net is drawn in so that the net becomes a bag or purse. A power winch drags the net alongside, and the fish in the purse are brailed into the vessel's hold.

Consider what this procedure can do if the net is set inside the line delineated by those triangular signs at a creek mouth. Inside that line the bottom has been raised by centuries of silting, a common situation at all creek and river mouths. In many, the depth of water is just about the depth of a seiner's net. Add another factor—fish arriving to go upstream and spawn commonly hang around the mouth until nature tells them it's time to get on with it. Such schools can be so full of fish that there seem to be more fish than water. A great temptation? It sure is. So you can see why the seiner was our chief concern. One such illegal set could wipe out a whole year's spawning run. Result: a blank year down the road for that creek.

FISHERIES POLICING STRATEGIES

Caught in the act (with his pants down?) meant an arrest, which meant paper work plus costs for all concerned and so on. Deterring infractions was much preferred, and I'm sure that the air patrol was especially effective in this respect, because of our relative speed and mobility.

The inspectors' boats were no faster than the fish boats, and some fish companies used speed boats to scout for fish and report findings to their own fishermen. I doubt they would neglect to mention seeing a patrol boat. It wouldn't take much of a navigator to figure out how long it would be before that patrol boat could arrive in their vicinity. We, on the other hand, could occasionally give a pretty good imitation of being two or more aircraft. Here is an example of one such day's irksome work.

We have circled a fleet of fishing boats several times at about 500 feet or so and then, obviously satisfied that no one is cheating, we break off and head north. Just far enough to be out of sight and sound, then into a lake we happen to know is full of hungry cutthroat. After an appropriate time, we leave the lake, circle back, and arrive over the same fleet from the south. We could almost hear it, "Now who the hell is that machine?"

This was work, mind you, for which in 1938, I was paid $130 per month plus $3 per flying hour. (Pause here while the less privileged eat their hearts out.)

Hard to believe that you can sneak up on someone with a Wright J6 racketing away, but I did once. We'd been cruising along at a few hundred feet, close to the timbered hillside to soak up as much noise as possible. We rounded a point and entered a bay and there is a seiner well inside the creek mouth boundary markers, hauling in his net. I land pronto and start taxiing toward the villain. It happened this day that we had on board not one inspector but two, and one had a warped sense of humour.

"Gordon," he says, "have you ever made an arrest?"

"Nope, never had occasion to."

"Well," says he, "you make this one, good practice for you."

Why I need practice, I don't understand—arresting isn't my bag; flying aeroplanes is. But, rather than appear chicken, I say, "Okay."

When we got close enough I switched off and coasted the last bit, climbed out the window and down onto the float, and tied the Bellanca to the seiner, all without looking up—expecting to get poked with a pike pole or worse. Then I step aboard his deck. That is when a shipowner is apt to do something nasty. Well, so help me, I am staring at six backs—the whole crew were so intent on hauling in their well-filled net and the power winch was making so much noise, they

The crew of a purse-seine, caught with their nets set well within the no-fishing boundary at Endhill Bay, British Columbia, is about to be arrested. *C. G. Ballentine*

hadn't heard us. So I had to walk over, decide that the seven-foot-tall guy is the skipper, tap him on the back and say, "You are under arrest." I don't think it came out as a squeak; but it was a great relief to see that he was as scared as I was.

We must have deterred a lot of potential poachers because arrests were quite rare. Penalties for those caught could be severe, especially for repeaters. But I do not recall experiencing any hard feelings, certainly none toward the air crew, and we spent many a pleasant hour loafing on the deck of some fish boat at anchor, until one or both of us had to go back to work.

THE PRICE OF ISOLATION AND OUR PHANTOM CRASH

What with variations on this fly fishing game, such as once around an island and back—no radio on board, so no position reports, no flight plans except for an inspector, nothing approaching a schedule—we could be pretty hard to keep track of. We seldom announced where we were headed next and probably weren't going there anyway. By no means did we fly daily—the total contract hours weren't enough— but when we did we could be sure every day would be different. No boredom on fish patrol.

All good tactics for my job, but tough on my family. When my father was dying in 1939, Canadian Airways Limited made strenuous efforts to find and replace me. But couldn't. I read of his death in a week-old paper picked up on a cannery dock.

On another occasion, it was tough on my wife and a whole lot of friends. Heading for Prince Rupert, I flew over Porcher Island to show Roy my old Vedette wreck, then began a long curving descent toward the harbour, passing over Lawyer Island off the mouth of the Skeena. At the oil dock I was helping Roy refuel when I was interrupted by a phone call from one of the local papers. "The captain of *CPS Charlotte* reports an aircraft going down in flames by Lawyer Island. Who was following you?" When the caller gave me times, I said that was when I passed the island—the captain probably saw the sun flashing off my prop. A few minutes later, a call from a second paper with the added report, "The aircraft was seen to hit the water." No report that the ship was on the way to the rescue—life was cheaper in those days? Then the Department of Transport and Lloyds phoned, so I said that we would go and take a look. Did so, and found nothing but one boat peacefully at anchor. Reported this to all four callers and forgot about it. But the papers did not, as we were to learn some time later.

Being the only aircraft regularly flying that coast at that time was one of the pleasures of our job; but it also meant that if an accident were reported, it was likely us. Walter Gilbert, our superintendent, gets the word in Vancouver that we are burned to a crisp, and about a week later when we arrive back home in Bella Bella my wife is looking at me kind of funny—she hasn't been able to understand why everyone has been so extra kind to her lately. As people should be to a new widow, of course.

EMERGENCY PRECAUTIONS AND PREPARATIONS

Had we ever been left stranded by a "mechanical" or weather, we were pretty well equipped to look after ourselves. Had to be, with no radio, no flight plans. On board we had Canadian Airways' standard kit for west coast summer, which could feed several people for several days—an axe, a 30/30 carbine, and sleeping bags. These were Yukon 90 x 90s, great in winter but a bit much for summer. We also had a bag of less Spartan edibles so we could have a meal wherever we happened to be. As well, we carried our personal gear, spare clothes, a .22 rifle, fly rods, and my office in a suitcase, complete with typewriter.

The time I really needed an emergency kit in 1928, it was such a ridiculous outfit that, ever after, I took considerable interest in what I had on board. The Vedette kit had consisted of an axe and a galvanized tin a cubic foot in size, soldered all around, completely air and water proof—ditto people proof. I used to tell people, "Of course it has a can opener; it's safe inside!" I expect we were meant to use the axe; but I have wondered, what if I had rescued the tin and the axe had burned up with the aircraft—which is what should have happened but didn't. Inside the tin: some chocolate, Oxo cubes, and a dozen ship's biscuits known as pilot bread. The "pilot" did not refer to us, but to the "iron men in wooden ships" of the old sailing days. A ship's biscuit is about four inches in diameter by an inch thick and cannot be eaten by twentieth-century man unless first battered with an axe or soaked for a day or so. Medical supplies were even less adequate. Two good items—a Boy Scout knife and a Marbles match case—I still use.

The feeling of freedom and independence that comes from being well equipped—self-sufficient—was further enhanced by the fact that we had no passenger responsibilities in the usual sense. The fisheries people were considered crew and didn't expect the TLC (tender loving care) we owed a paying passenger. None of us went looking for discomfort, but if a little did find us, it wasn't important. Add to this an air engineer/mate on board, and a pilot's life became rosy indeed.

COPING WITH BC COASTAL WEATHER AND WILDLIFE

It wasn't always sunny—it's not called the "rain coast" for nothing. The annual rainfall at City Hall, Vancouver, is about 55 inches, more as you near the mountains, less at the airport. Swanson Bay averages some 200 inches and in 1939 managed to record 255.

What this amount of precip does to buildings, when it falls as snow, is smash them flat. In 1928 the town and mill had not been abandoned very long, and the site was intact: rows of houses, a hospital, and a three-storey hotel. By 1939, the hotel was a pile of kindling with a roof on top, and of the houses, only the caretaker's was habitable.

What the rain did to us was no problem. We wore old clothes, a long slicker with a sou'wester at the top and hip waders at the bottom. What it did to operations was cancel them if too bad, or require some pretty low flying, sometimes just off the water. Now a wise man does not eat the snow when it is yellow, does not play leap frog with a unicorn, and does not fly at one foot in unsafe conditions of wind and water. But given that, it can be quite safe; after all, it's just the penultimate stage of a landing. Sit there with one hand on throttle, one on stick, and when things get too bad, just ease off the power.

One day in 1939, Roy and I went one stage further—we taxied some 60 miles. We left Rupert with one fisheries man; he headed for Butedale, we for Swanson Bay. Flying down Grenville Channel the ceiling kept lowering; finally an extra thick squall put us on the water. We had passed Lowe Inlet, where we could have sheltered, but none of us wanted to back track. Decided to taxi, figuring the weather had to get better. As always, it did, but not that day. We chugged along, once in a while able to get up on the step for a bit, on down Grenville Channel, across Wright Sound and into McKay Reach—Roy and I took turns with the rudder so the other could play cribbage with the passenger, managing to take his money. As we neared Kingcome Point, it was obvious that it would be black dark before we could reach Butedale. I remembered an abandoned logging outfit a little way up Ursula Channel in Goat Harbour—with a dock we could tie to and a shed for shelter. So there we went for the night, then on to Butedale for breakfast next morning.

This was my second long taxi trip. In 1928, I headed home from the Charlottes in a Boeing B-IE—Bill Upham (pilot), Alf Walker (air engineer), and me. We were forced down in Otter Pass by darkness. We taxied some 95 miles to Swanson Bay in the dark, once nearly colliding with a whale. In those days, and maybe still, places you could moor an aeroplane were scarce, so playing the steamboat was the answer. Did no harm as long as the revs were adjusted so the prop didn't pick up spray and damage itself.

Our freedom to fly much as we pleased most of the time—or to taxi—gave us lots of opportunity to see wildlife of all sorts. Whales and basking sharks were common sights—on one occasion very close indeed. Bill and I were lazily unloading and securing 'BFC; we had just landed from a patrol on a beautiful, calm evening. A whale was cruising Swanson Bay shoreline, down for a gulp of food and then up for a look or a breath. As we watched, it became apparent that one of the ups would be right in front of the float on which we were standing. So I opened my camera, measured the light and set it for distance, and placed my finger on the cocked shutter release. All set, eh? Sure enough, we got an up, right in front of us, not 10 feet distant. A very wide mouth could have held Jonah seated on a stool (And I guess, if you were in a whale's mouth, there would likely be a stool or two?). I would have included the photographic proof with this story except for one thing—which I am too embarrassed to mention.

Lots of eagles—one we often passed had a nest atop a tall dead tree. Didn't seem to mind as long as we passed; but the day we decided to circle for a closer look, he/she came up after us. We left. I hit a small duck once—smashed the windshield and caused a forced landing. Eagles were out of my league entirely.

Other airborne hazards were the "no-see-um," an extremely irritating mini-mosquito, and the horsefly. Horseflies don't sting, they bite—hard and deep—and it hurts. We used dope (repellent) for both pests, the strongest we could get. I don't remember the name, but I do recall one day when we'd taken some flies up with us—Bill and I got out the dope bottle and applied it liberally. While my hands were still

Against a spectacular mountain backdrop, pilot Gord Ballentine (standing), engineer Roy MacDougal (hatless), and passengers stop for lunch in the summer of 1939. Behind them is a Pacemaker, CF-BFB.
C. G. Ballentine

wet, I began to write up the log, whereupon my pre-Bic fountain pen went as limp as cooked spaghetti, became a Salvador Dali master-piece. Potent dope obviously, but more lethal for pens than horseflies.

Wolves we saw only during the spawning job and only on Banks Island. We watched a goat race across the snowfield above a lake once. Bears were more common—one favourite lived in a cave about 500 feet up a hillside we often passed. If he were home at the time, he'd stand up and make like King Kong. Seemed to think I had Fay Wray sitting alongside me. How myopic and wrong he was—no way did Bill Jacquot resemble that lady.

One day we saw an *Ursus Kermodei*, a rare off-white bear first spotted on Princess Royal Island—where we saw ours—from which it has spread.

MAKING DO IN THE WILDERNESS

Because we'd be far from maintenance or overhaul facilities for the duration of the contract—four or five months, our aircraft out in the weather at all times—it made sense to simplify things as much as possible, electrics in particular being vulnerable to salt air damage. So no radio—it didn't perform so far from home, and anyway we didn't want fishermen listening to any position reports. No electric windshield wiper; an old manual was fitted and worked fine. The battery was removed, which meant hand-cranking the inertia starter, and, to simplify the operation, we bolted the crank handle to the starter shaft.

Regarding the crank-handle modification—when we got home in the late fall and back on schedules and charters, our chief mechanic, Bill Tall, insisted the handle be unbolted. Looked untidy, he said. (On an old Pacemaker?) I pleaded, to no avail. So one day, I'm on my own; I take a couple of prospectors into a lake, they disappear into the bush, I shove off from the beach, get out the handle (greasy), begin to fit it (hands sweaty), and you guessed it. Into 20 feet of gin-clear water.

There must be lots of old inertia-starter hand-crankers out there—did any of you ever do it with an ordinary screwdriver, crosswise in the notches? I managed to start twice (Didn't have much choice, did I?), once to get out of the lake and again at a refuelling stop on the way home. Bent the screwdriver.

Despite being out in the weather for all those months, the Pacemaker gave us no serious trouble, partly, I am sure because, in Bill Jacquot, I had a first-class engineer. The engine mount had to be welded twice in 1938, once in Rupert, another time at Ocean Falls, where Bill could borrow equipment. The Wright J6 had an occasional tendency to suddenly indicate zero oil pressure, so we'd land and flush out the pressure relief valve. It always happened where a landing was no problem. Salt water barnacles on the floats and salt water corrosion were a threat everywhere that we tried to minimize whenever we got into a fresh water lake. And in 1939 when the RCAF anchored a scow in Gulichuck near Bella Bella, with workshop and ramp, the troops let us use their facility a couple of times.

Fuel: aircraft were so rare on the north coast it was no wonder oil companies were in no rush to establish stations. A few places—Bella Bella, Ocean Falls, Prince Rupert—had facilities; elsewhere we did for ourselves. Imperial Oil, big in aviation products in those days, and Union Steamships both gave us good service, dropping off drums when requested. These were cached along with a wobble pump and funnel.

A typical British Columbia mountainscape, above the head of Dean Channel, to the right of Kimsquit Lake: a magnificent sight—as long as the engine kept running. *C. G. Ballentine*

THE END-OF-SEASON SALMON COUNT

Finally the end of the fishing—no more cops and robbers—now our job is to use the hours remaining in the contract to help with the count of spawning sockeye salmon.

Sockeye are fussy about where they spawn. They want a stream that flows into a lake. Inspectors used to speak of them as "ripening" in the lake before going upstream to spawn. They want to count the number spawning in the same half-mile or whatever, as nearly as possible on the same date each year, so that an annual comparison is reasonably accurate.

To do the count can mean hiking up the stream from the salt chuck, going around the lake (or making a crude raft), then hiking some more up to the spawning-count area. And if you have ever tried hiking up a wilderness creek in British Columbia's coastal rain forest, you know it isn't a stroll in the park. Rocks, deadfalls, jungle-thick undergrowth, likely some devil's club as well. Tough going. In some creeks it has to be done the hard way, but never by choice, and that's where we come in. If the lake is large enough and the weather fit, we'd fly them in and put them ashore at the mouth of the spawning creek.

This work got us into beautiful lakes no white man would ever visit in those days: Ellerslie, Ingram, Msameet, Sagar, Kitkiata, Howyet, Kwakustis, Tankeeah, Long, Weare, Curtis, Mink Trap, Cridge, Quinstonsta, Deer, Mikado, Gale, Bare, Kimsquit, Walker, Koeye, Mooto, Klewnuggsit, Namu, and Owekano.

With the new pleasures came some new problems and pressures. Say we were supposed to do Lake "X" on such a date, same date as last year. We need better weather than we need for patrolling, but it is late in the year and the weather is worse overall. We want to get the job done and go home, we've been many months away—now we must guard against that urge leading us into stupidity. Just the same, the lake work was a very interesting part of the season's work.

Some inspectors liked us to walk shotgun with them because spawning salmon are of interest to bears as well as fisheries people. Mildly adventurous, carrying a 30/30 up a creek in virgin country—I've listened to grizzlies roaring not far off, but never met one close enough to worry either of us—if anything worries grizzlies.

Wolves were uncommon, but we saw them on Banks Island. Big, blue-black animals. This was the first time that an area inspector had a sighting, and direct evidence of wolves feeding on spawning salmon just like bears. They take one large bite, for which they are well equipped, just back of the head where there is most of the oil and meat, then flip the carcass onto the bank and go for another.

Do salmon enjoy sex? Well, if they don't, after fleeing from poachers, bears, wolves, eagles, and ravens, struggling up waterfalls, no food since leaving the sea—if they don't, life is indeed unfair. And then having to do it the only way they can—well.

We had put Charlie Lord ashore at the head of Long Lake. Charlie was demonstrably the best crib player between Cape Caution and Rupert—wore Coke-bottle glasses—and was the hardest working inspector I knew. Legend had it he was a tough union man, always figured the Department had hired him just so they could fire him and get even. Charlie was giving no one a chance to do that. As usual I ask, does he want me to come along with a gun. Charlie says, "Naw, who

Fisheries inspector Gordon Reade on the float of the Bellanca after a strenuous hike up Koeye River, British Columbia, to conduct a spawning inspection. *C. G. Ballentine*

needs it?" (Next year he needs it but doesn't have it—he and a bear spend half an hour on opposite sides of a tree until Bruin, shocked by Charlie's language, goes away.)

Anyway, after a while I get bored; Roy is snoozing, so I grab the 30/30 and start up the trail. Pleasant river, about 50 feet across, very shallow now. Sit on a rock and idly watch the swarming spawners. Salmon do it this way: the doe scrubs a small trench in the gravel with her belly—the buck is finning in the current just ahead—she squirts out some eggs, he dribbles out some milt, which they hope will drift down and contact some eggs. Doesn't look very efficient or like much fun.

No such action in front of me at the moment, but a doe is busily nudging a buck where it will do the most good. He looks bored, obviously not his first today. Close by is another doe (they outnumber bucks that year) behaving rather twitchy. And the instant Lady Number One gets what she wants, over comes Lady Number Two, shoves Number One out of the way, and goes after that poor glassy eyed buck.

There is the evidence—does it prove enjoyment? Or is it a case of "If she's getting it, I want some too"?

OUR OLD BELLANCA FOR A NEW WACO

A Waco "C" (cabin biplane) came into our lives during the 1938 spawning season. Bill and I had been loafing in Ocean Falls for a couple of days waiting for weather fit to take some inspectors into Kimsquit Lake. Howard Macdonald arrived in the Waco, CF-AWL, with orders to exchange it for our Pacemaker. Neither we nor the fisheries people were happy about this; we liked the Pacemaker—but *orders is orders.*

As I had never even sat in a Waco, I decided I'd better get to it, so Bill and I start up and taxi out into Cousins Inlet. Feels comfortable, controls and gauges all seem to be as they should be, so off we go. Takeoff is prompt (we're empty); all is fine. Until, at 50 feet or so the Jacobs' roar suddenly changes to a feeble sputter. We land. No problem, this thing handles very nicely. We look around, can't see anything wrong, start up and do it again. All of it—for landing number two. After takeoff and landing number three, I decide I am qualified on the type. I also decide that an aircraft with a service ceiling of 50 feet, that provides three forced landings in 15 minutes, is not capable of flying me 35 dry miles to a mountain lake.

So next day when the weather improved I flew the Pacemaker into Kimsquit, leaving Bill behind to work on the Jacobs (engine). From Ocean Falls to Kimsquit Lake you fly Dean Channel, passing the cliff-side legend, "Alexander Mackenzie from Canada by land, the twenty second of July. One thousand seven hundred and ninety three," a gentle reminder that we should give thought before claiming to be pioneers. Turn left and follow the river. Small river, okay as a route marker, but not for a forced landing.

Kimsquit Lake in the spawning season is a spectacular sight. It sits in a bowl in the mountains, a lovely slate blue from glacial silt. When filled with sockeye—which at this stage in their life become a bright red—the result is a magnificent goldfish bowl.

By the time I got back to Ocean Falls, Bill had 'AWL's Jacobs in good shape—problem had been with the half-battery/half-magneto system. So we accepted the swap, transferred all of our gear from the Pacemaker, and set off for Swanson Bay.

Early on I had said that on day one of 1938 I wasn't very happy about leaving my wife behind. I should also admit to just a bit of apprehension. Some of my friends who had been on fisheries before me were not above embroidering the hazards and hardships—bullying inspectors, frightful weather to fly in, the time Bill Lawson had to half-loop, half-roll (an Immelmann?) a flying boat to escape from a dead-end valley, all that sort of stuff. I was skeptical. I'd been flying the coast for a few years and knew better. But not 100 percent. After all, in my first season, 1928, two aircraft crashed, and the HS-2L was destroyed when its Liberty engine blew up—one man was killed, another badly injured. Enough for an active imagination to prepare me for the worst.

Absolutely none of which came to pass. Never did anyone question my judgement or decision when safety was involved. The only time an inspector got mad at me about weather was in reverse, so to speak. Roy and I took off alone to patrol in weather Gordon Reade thought was unsafe. Happy days, eh?

But I did have two bad scares in two years. Statistically that is not so bad. Statistically. Read on.

En route back to home base, everything worked nicely, we had

Inspector Gordon Reade and air engineer Bill Jacquot about to board Gord Ballentine's Waco Standard, CF-AWL, at the unnamed second lake above Kis-A-Meet Lake, British Columbia. *C. G. Ballentine*

both begun to like the Waco. Comfortable, very nice to handle, very responsive compared to the Pacemaker. Which led me to suggest, at about 3,000 feet, "Let's see how tight a turn this thing will do." Okay by Bill, so around we go. During the turn there is a rumble from aft. Surprised us but that was all, until I rolled out of the turn and could hardly get the nose down with the (control) yoke right up against the panel. Long before we get to the deck I had to get Bill to add his push to the yoke. And our combined strength was just enough to last until we landed, both near muscular exhaustion. Very good thing I hadn't started this nonsense higher than 3,000 feet.

Turned out the problem that gave us bad scare number one was the tailplane adjustment mechanism—control wire wrapped around a fibre drum, but not tightly enough. The hard turn had caused the drum to spin past all safe limits to full nose-up. When Bill fixed this, our mechanical problems with 'AWL came to an end.

A few days later I was asked to take two inspectors into Mikado Lake. No pressure. They even proffered the information that in previous years some pilots had refused the trip. Some had gone in, and, more important, gotten out. I looked it over very carefully, small lake down in a basin surrounded by trees, no sign of strong winds, and I decided that if I could get unstuck before passing a certain rock, take-off would be safe enough. So in we go, and the inspectors do their spawning count. When it's time to leave there is still no sign of wind pouring down into the lake (that stuff—wind—should be coloured red or yellow), but I decide to be extra careful anyway. Taxi to windward end, get on the step going downwind, skitter around into wind at extreme end, and we are off the water—well before my go-no-go rock—and climbing nicely.

Too soon to relax, but we are higher than the tall timbers ahead. Then bingo into subsiding air, pouring over the trees and down into the bowl like a waterfall—and no way does the Jacobs have the power to paddle up the waterfall. And now no room to turn with an ugly old dead tree straight ahead and higher. Well, with no option, I rolled the left wing over its tip with, I swear, no more than six inches of air. That was a very quiet lot of people for a while, as we gradually recovered from scare number two.

One of my regrets is that I never made anything but the briefest notes in my logs. Times, places, no remarks. The above incident is noted with "whew," nothing else. Next year, I did that lake again; no remarks, so presume it was routine.

There was an embarrassing sequel to the bad scare. About a week later I was asked to go into another small lake. After looking it over I said sure, provided we position a fisheries boat nearby with an empty drum into which we can off-load most of my fuel. Well, when I came out of that lake I must have had 500 feet to spare.

Today's seaplane pilots who may be surprised that we thought certain lakes too small, or marginal, should bear in mind the differences in aircraft. When I was with PWA in the late 1950s, having been away from floats for years, our VFR Division's operation with Beavers and Cessnas was quite an eye opener to me.

When there are only four hours left in the contract, saved to get us home, our work is all done. Matter of fact, with my usual good luck, it really is done. Completed, both seasons. We load up all the personal

The author's Bellanca, resting on an unnamed lake, is almost lost against a massive mountain flank. Here, the aircraft crew and a fisheries inspector caught 14 cutthroat trout in 20 minutes.
C. G. Ballentine

and company stuff we can carry, then crate and ship the rest. Leave gas caches for needy charter flights—or for the next season's patrol—which after 1939 was cancelled because of the war.

If I haven't created a little envy among seaplane pilots, I must need a new typewriter.

A Bit of a Dog!

The Fairchild KR-34 Was Underpowered for Float Operations

JIM TAYLOR

Compared with cars and trucks, aeroplanes are relatively long-lived. A few Douglas DC-3s remain active commercially more than six decades after the first flight of the prototype. But, while a working aircraft's performance may have been more than adequate when new, this may no longer be the case should it soldier on to compete with a later generation of comparable machines. A pilot with expectations attuned to current types would find its performance less than sparkling.

Such was the case in 1938 with Jim Taylor and the Fairchild KR-34 owned by the Ontario Provincial Air Service (OPAS). The small radial-engine biplane had been designed 10 years earlier in the U.S. to appeal to the well-to-do sportsman pilot of that era. Although certified for use on floats and skis, it was intended primarily for recreational flying on wheels from grass strips. As a seaplane it was a marginal performer. The manufacturer likely never envisaged the demands that would be made of this machine as a working aircraft in the Ontario wilderness.

The OPAS acquired only a single KR-34C, probably to test it as an alternative to the D.H. Gipsy Moths that had served well in the fire-spotting role. However, when observation and supply duties were combined in a single aircraft, the Stinson Reliant, the small biplanes became redundant and were sold to private owners.

CF-AOH, as the KR-34 was registered, eventually became the property of Sonny Dale. In August 1947 Sonny ran the aeroplane into

111

the trees on the shores of Wildcat Lake, Ontario, and abandoned it there. In 1963 the Department of Lands and Forests recovered the remains, though it was 30 years before the plane was restored. The KR-34, now fully airworthy, can be seen in the Canadian Bushplane Heritage Museum at Sault Ste. Marie, Ontario.

Jim Taylor joined the OPAS as an apprentice in 1935 at age 21 and was sent solo by the legendary George Phillips, an early OPAS pilot, in 1936. As well as the KR-34, Jim flew the Service's D.H. 60 Gipsy Moths and was checked out on the Reliant in 1939 by OPAS Director George Ponsford. He remained with the Service under its various names until his retirement in 1973. Jim's last aircraft, which he flew out of Toronto Island as director of personnel, was the Beaver CF-OBS, "Oscar, Bravo, Sierra," now exhibited with the KR-34 in the CBH Museum. His story appeared in the Spring 1996 CAHS *Journal*.

The Fairchild KR-34 had two open cockpits. The pilot sat in the rear cockpit, and two passengers sitting side-by-side could occupy the front. An R540 Wright J6 Whirlwind, rated at 165 hp, powered the aircraft. Frank MacDougall, superintendent of Ontario's Algonquin Park (later deputy minister of Lands and Forests) piloted this plane until the start of the 1938 season, when a Stinson SR-9FM Reliant replaced the Fairchild. The Stinson, with its P&W Wasp Jr. engine, developed about 460 hp at full power and cruised on about 360 hp. Frank must have found this a pleasant change after the KR-34, which was a bit of a dog.

I inherited the little Fairchild from Frank and can speak from experience. Apart from performance limitations, of which I will say more later, there were other shortcomings.

For one thing, starting the KR-34 could be extremely aggravating. Because there was not enough space to stand on the float behind the propeller, it was not practical to start the aircraft by swinging the propeller. Instead, you made use of a crank and an inertia starter. You inserted a hand crank into a shaft that protruded from the cowl just behind the engine. The shaft was connected with the starter, which, when cranked, started a series of encased, heavily weighted wheels revolving rapidly. When the sound of the starter reached a high whine—or if you got the feeling that it was about to fly apart through centrifugal force—it was time to transfer its energy to the engine.

When a helper was available, starting from a dock was no problem. But many starts had to take place when I was alone on some remote lake. So a start went something like this. I paddled the aeroplane out a short distance from shore. Then I hopped into the cockpit, primed the engine, and, with the throttle set, crawled back out onto the lower left wing and inserted the crank. I wound up the starter and, when the sound was right, scrambled back into the cockpit and engaged the starter with the engine—at the same time cranking frantically on the magneto coil. If everything had been set properly, I got a start. If not, there was no more sickening sound than that of the starter rapidly losing its energy without the expected noise of the engine kicking in—meaning that I would have to repeat the whole process.

But by now another insidious factor might affect the procedure. All this while the aircraft could have been drifting with the wind. I could be back in the cockpit—all ready for a start—when I would look up to see that I was facing the shore. I would have to abort the

The Fairchild KR-34 at the Sault Ste. Marie, Ontario, OPAS headquarters. CF-AOH was painted in the Service's livery of aluminum and yellow with black trim and registration. Fully restored, it is now on display at the Canadian Bushplane Heritage Museum in the Soo. *DND RE 64l3252*

start, reorient the aircraft with my paddle, and start over again.

One very troublesome day, my attention had been totally taken up with the start. When I finally did manage one, I saw that I was about five feet from a shoreline heavily treed with cedar. In sheer exasperation, I rammed the throttle forward. But before I could turn off the switches, the whirling propeller digested about four bushels of cedar boughs and distributed them throughout the engine and open cockpits.

In the spring of 1938 the ministry agreed to allocate an aircraft to Temagami for fire protection purposes, and I was told to report with the Fairchild. Hap Bliss, the Temagami chief ranger, kindly provided a small log cabin on Forestry Island for air engineer Ernie Drew and me. While I was not aware of it when I arrived, some of the ground staff were skeptical of the usefulness of aircraft. I suspect this was the result of disillusionment with services provided in earlier times, or perhaps a lack of understanding of the limitations and capabilities of an aeroplane.

There was no problem with Hap, though. He appreciated aircraft and was most anxious that the number of hours flown should justify any pitch that may have been needed to get the KR-34 there in the first place. And in retrospect, perhaps an aircraft sitting in front of his headquarters lent a bit of prestige. In any case, even though it was a light fire year, Hap kept us busy on general transportation and detection. My big selling problem was with Phil Hanson, the chief ranger at Latchford. Phil was a dour individual who didn't have much use for aircraft in his division. But, after a bit of pestering on my part, he agreed to let me at least show him what the machine and I could do.

He asked me if I could land on the small lake near the Maple Mountain tower. When I asked him if it had been used before, he replied that another OPAS pilot, Vern Gillard, had landed on it in a Gipsy Moth. I flew over and had a look—the lake sure seemed small. But with my strong desire to please, lack of experience, and knowing that Gillard had landed there, I gave it a try. Although there wasn't much room to spare, I found that I could get in and out. I serviced the Maple Mountain tower through the spring and summer without incident.

Earlier, I described the KR-34 as a bit of a dog. The maximum rpm its Wright could get on takeoff was around 1,600—not enough to

Jim Taylor with the Fairchild KR-34, CF-AOH, of the Ontario
Provincial Air Service, which he flew in 1938. The location is the
OPAS's Temagami, Ontario, base. *J. Taylor*

develop full horsepower. Consequently, as soon as the Fairchild was airborne, you put the nose down, allowing the engine to rev up and develop the speed and additional hp needed to climb. This procedure was no problem on large lakes, but it nearly cost me my life when I returned in the fall to the small lake at Maple Mountain.

I went in without paying much attention. But, after unloading cargo destined for that stop, I realized that the lake was even smaller than I remembered—summer evaporation and lack of rain had significantly reduced its size. Worse, the wind that had aided my landing had disappeared. Calm, glassy water can make taking off difficult—there are no waves to break the water suction under the floats, to let you get on the step. So the takeoff run becomes much longer. There is no problem if you have lots of power or water; I had neither. Nor did I have the patience to wait for a good stiff wind to stir up some waves.

So I started taxiing around the circumference of the lake. This served two purposes: it created waves and allowed me to pick up speed for a straight takeoff run in the direction of the shoreline where the trees were the lowest. When I finally turned into my run I became airborne about halfway across. But I soon realized that I did not have enough air speed to climb—and I had reached the point of no return. I did not have the speed or the room to turn. And no place to land except in the trees!

I immediately pushed the stick forward and headed straight and level for the shoreline. At the last moment, I pulled back, hoping that I had gained enough speed to clear the treeline. At the peak of the climb, with speed falling off and a stall imminent, I pushed the stick forward, dropping into level flight and struggling for flying speed. Just as I did this a cardboard box of bread on top of the cargo in the front cockpit became weightless and gently drifted out into space. As it floated by I put out a hand and pushed it clear of the tailplane. I can't remember how long I struggled along just over the trees at two mph above stalling speed. Eventually the topography dropped off sufficiently to allow me to put the nose down and pick up safe flying speed.

When I got back to base, I called Phil and said, "The Lord gave me a bit of a push to get out of that lake at Maple Mountain. I don't think He wants me to go back in—and I am going to pay attention to Him." With that, Phil laughed and told me how Gillard had used the lake.

He had force-landed there and, before attempting to take off, had stripped his Gipsy Moth of all excess weight. He reduced the fuel load to an absolute minimum. Then he had tied the tail of the machine to a tree and, after reaching full rpm, had a helper cut the rope—which put him into the air in short order.

Phil never asked me to use that lake again.

The last aircraft flown by Jim Taylor was de Havilland Canada DHC-2 Beaver, CF-OBS, seen here at the Science North (museum) dock at Sudbury, Ontario, in August 1987. Like the KR-34, this machine is also on display at the Canadian Bushplane Heritage Museum in the Soo. It was the second Beaver built. *W. Wheeler*

The Clipped-wing Anson Episode

A Daring and Ingenious Improvisation

T. P. (TOMMY) FOX

Asked to name a Canadian airline owner who has risen from meagre beginnings through determination and entrepreneurial skill, most Canadians would probably answer Max Ward, or possibly Grant McConachie. Few would think of the author of this account, Tommy Fox. Yet he was an equally worthy if little-known member of this select group. During the mid-1950s, Fox's Associated Airways was the principal Distant Early Warning (DEW) Line contractor for western Canada, employing 200 people and operating a diverse fleet of aircraft as well as subcontracting to such major firms as Pacific Western and Queen Charlotte Airlines.

Born in Vancouver in 1909, Tommy Fox learned to fly in 1929 at the Sprott-Shaw School of Aviation on a Waco 10, open-cockpit biplane. In 1932, he and a partner built a Pietenpol Air Camper, registered as CF-ATU, in order to make their flying more affordable. Plans for this very basic, open-cockpit parasol with tandem seating for two could be purchased in the U.S. Home-building was then almost as popular as it is today. Nearly three dozen Pietenpols would be registered in prewar Canada. To power 'ATU, Fox and his partner salvaged an air-cooled Cirrus aircraft engine—probably from a written-off Moth—instead of using the more common modified Ford Model A automobile engine.

During World War II, Tommy Fox flew Avro Anson V twin-engine trainers as a staff pilot with No. 2 Air Observer School (AOS) at

Edmonton, Alberta. This extensive Anson experience, as the following account demonstrates, would later stand him in very good stead. From the AOS he transferred to Ferry Command, where he completed 16 Atlantic crossings delivering North American B-25 Mitchells and Consolidated B-24 Liberators to England. His final RCAF posting was with Transport Command. He begins his story at this point, immediately upon his departure from the Service.

Following the satisfactory resolution of the incident in question, his company continued to expand. Fox also became the local distributor for Taylorcraft, an American builder of popular light aeroplanes. Along with the D.H. 90 Dragonfly and the Anson of the story, Associated Airways' fleet would include four other Anson Vs (purchased from the War Assets corporation for $2,500 each). As well, Associated owned a Cessna Crane ($500 from War Assets), a pair of Barkley-Grow T-8P1s (small twin-engine airliners), a Noorduyn Norseman, a Bellanca Skyrocket, two Helio Couriers (flown for the Alberta Government), a pair of DHC Beavers, and an Otter. Associated's larger aircraft included a Lockheed 14 and, for DEW Line work, a Curtiss C-46, and a Bristol Freighter (the first of its kind in Canada).

In 1950, Tommy Fox purchased his first helicopter and founded Associated Helicopters. Six years later, a pair of serious accidents led to the sale of Associated to Pacific Western Airlines with Fox becoming vice-president of the expanded company, in charge of Northern Operations. He retained ownership of Associated Helicopters. In 1976 he retired from aviation to raise horses and cattle on his ranch near Edmonton, Alberta, where he resided until his death on 14 September 1995. His story appeared in the Summer 1990 issue of the CAHS *Journal*.

In May of 1945, I obtained my release from 45 Group RAF Transport Command and ferried an ex-RCAF de Havilland Dragonfly from Montreal to Edmonton with my wife and children on board. I had purchased this aircraft, assigned registration CF-BZA, from Crown Assets earlier in the year and worked on it between trips to obtain its Certificate of Airworthiness. I also purchased a Tiger Moth, CF-BEN, located in Neepawa, Manitoba. After arriving in Edmonton and obtaining a temporary place to stay, I picked up the Tiger Moth.

These two aircraft formed the initial fleet of Associated Airways Limited. In the beginning, Associated's activities consisted mainly of passenger hopping, with the occasional charter flight.

At this time the company was mainly a one-person operation—I was pilot, engineer, janitor, and bookkeeper. My wife assisted by selling tickets, loading passengers, and running errands. Later that summer, we started a flying school, and I hired Art Bell as instructor. Art had the distinction of being Associated's first paid employee. Receiving a previously applied-for licence from the Air Transport Board, I was able to expand our operation. Associated would enjoy steady growth, and its fleet and staff increased accordingly.

AIRCRAFT DOWN

In the early years, especially, Associated Airways flew many so-called mercy flights, including one "rescue" flight of a somewhat different nature. One of the company's pilots, delivering supplies from Edmonton to the Hudson's Bay post at Chipewyan Lake, Alberta, in a ski-equipped Anson, CF-EKD, decided to fly above the cloud layer. This was strictly against company rules, not to mention Department of Transport regulations. Finding no convenient holes under him at his estimated time of arrival over his destination, he let down through the clouds, loading the wings with ice in the process. When he levelled out to land, the ice caused the starboard wing to stall and strike the surface of the lake, breaking off some nine and one-half feet from the tip. Despite this, he was able to taxi up to the Hudson's Bay post and unload his cargo.

On the following day, another pilot flying south noticed the damaged Anson. He landed and brought our unfortunate pilot back to Edmonton. When he arrived at the Associated hangar, we questioned him about the circumstances leading up to the accident and, particularly, on the extent of the damage. Associated had only been in operation for about a year and a half, and the loss of the Anson would be a severe blow. There were no roads into Chipewyan Lake; if the ship were to be recovered, it would have to be flown out before spring break-up.

After careful calculation, I was able to establish what the wing-loading per square foot would be for the empty aircraft, if we squared off the damaged wing and cut a corresponding nine and one-half feet

Work has just begun on the damaged starboard wing of CF-EKD at Chipewyan Lake in northern Alberta, January 1947. *T. Fox*

from the opposite wing. This compared reasonably well with the wing-loading when maximum weight of freight was carried. Flying the ship out definitely seemed feasible.

We decided to "rescue" the Anson, and I took off in the Taylorcraft, CF-DBY, accompanied by Maury Danes, with tools, fabric, and dope. We arrived at Chipewyan Lake at 16:30 on 25 January 1947. The post was not hard to find since it was the only building on the lake, or anywhere in the area, for that matter. Upon landing we taxied up and were greeted by the post manager and his wife, who welcomed us with true northern hospitality, offering the use of their spare bedroom and of course to share their table at meal times.

Then Maury and I went out to the Anson for a first-hand view of the damage. To our dismay, we found it to be all the pilot had described and more. Not only was the outboard half of the aileron shattered, so was the rear inboard of the aileron hinge. Obviously, if the ship were to be flown, the spar would have to be extended far enough to accommodate the hinge. This was a woodworking job, and we had no material with us. A search of the post revealed that the only lumber available was a few scrap ends of maple flooring left over from

121

when the place was built. So maple flooring would have to take the place of the usual Sitka spruce.

In due course, the wing and aileron were squared off, the rear spar extended, the aileron hinge re-installed, and the stub end covered with fabric and doped. Our next job was to saw off the other wing in similar fashion and to cover and dope it. During the several days it took to effect these repairs, the temperature stood at -50 (F) and colder. The wing and engine covers served as a makeshift shelter with two blow pots to raise the temperature a little.

While we were there, a pilot flying an Anson for another company found that, due to headwinds, he would not have enough fuel to make Edmonton. He landed on the lake and asked if he could borrow some gas, assuring us that he would return the fuel on his northbound trip the next day. We were agreeable, and he pumped about 45 gallons out of the starboard wing tanks. The following day, he returned with a drum of gas, and, because we were working on the starboard wing tip and the blow pots were burning, the pilot pumped the gas into the port wing tanks. This, as it turned out, would cause some unexpected problems on the ferry flight home.

By 30 January, the wing "repairs" were completed. It was time to make the appropriate entries in the aircraft logbook detailing the work we had done. Normally, I would also have to certify the aircraft as airworthy. But knowing the Department of Transport officials in Edmonton would be less than pleased with the Anson being flown in its new configuration, I signed the logbooks certifying the aircraft as "not airworthy" and "not fit for flight." This was a ploy to safeguard my air engineer's licence from suspension; I was the only one with Associated holding such a licence at the time. Since it was mid-afternoon and too late to start the ferry flight home, we decided to heat up and start the engines for some test runs.

I taxied down the lake, turned into the wind, applied power, and ran for about a mile without trying to lift off. On a second run, I lifted the ship off the snow about two feet to get the feel of the aircraft in its modified configuration. On the third run, I took off and did a circuit. The ship lifted off at 90 mph. From our tests, I found that it took a lot of power and speed to take off and that cruising speed was slow even at climb power. Also, with only half of the normal aileron area, and that being inboard, lateral control was slow and sloppy. I landed and

Tommy Fox (left) and Maury Danes at Chipewyan Lake, Alberta, with CF-EKD, 1947. *T. Fox*

taxied back to the post, where we diluted the engine oil, loaded bits and pieces, and made plans for departure early the following day, 31 January.

The next morning, after breakfast with the post manager and his wife, we fired up the blow pots to preheat the engines on both the Anson and the Taylorcraft. With the temperature at -34 (F), this took some time. We thanked our hosts and bid them farewell before starting the engines in our aircraft. When normal cylinder-head temperatures had been reached and the usual preflight checks made, we taxied out for takeoff using the ski tracks from the previous day's flights. I took off first in the Anson, followed by Danes in the T-craft. After circling the lake to gain altitude and a cautious waggle of the wings over the Hudson's Bay post, I headed the Anson southeast toward the bend in the Athabasca River that would lead to Fort McMurray. The Taylorcraft set course south for Edmonton. Takeoff time was 13:15.

I maintained a gradual climb to an altitude of approximately 2,000 feet. The fuel gauge for the starboard tank seemed to be dropping at an alarming rate, undoubtedly due to my having to cruise at climb power just to maintain 90-mph airspeed. I turned on the fuel tank cross-feed valve to enable the starboard engine to draw from the port tank; however, after a few seconds, the engine started to sputter from lack of fuel and I had to switch back to the starboard tank. I tested the cross-feed several more times, always with the same result. Obviously, it was frozen. Although the bend in the Athabasca where it heads east to Fort McMurray was just visible on the horizon, I doubted whether there was enough fuel in the starboard tank to keep the engine running until we reached that settlement. Evermore frequent glances at the starboard gauge showed it rapidly approaching the empty mark. The thought of attempting to fly 'EKD in its unusual configuration on only one engine was anything but pleasant. I would have to force-land.

A few miles beyond the bend in the river, while I was flying directly over it toward McMurray, the engine sputtered to a stop. I feathered the starboard propeller and reduced power on the port engine. Then I attempted to reach the radio station at McMurray or any radio station within range on a mayday call—but without success. All the while, I maintained a gradual descent into the river valley, which, at this point, was winding, with steep 500 to 700 foot banks on either side. My approach had to be straight ahead following the bends in the river. I had one more turn to negotiate before reaching the point of having to land. I made it and faced a straight stretch of a half-mile or so. This was it! I reduced power and set her down smoothly, completing my landing run close to the north bank. Before shutting down the port engine, I attempted once more to reach someone on the set, but was again unsuccessful. So I made a blind radio report on my problem and location before diluting the engine oil and shutting down. The time was 14:25.

COLD WEATHER SURVIVAL

After putting the covers on both engines, I mentally reviewed my situation. Since my location was only about 15 miles west of McMurray, I decided to strap on the snowshoes and walk out. I left a note in the aircraft outlining my plan and gathered up an axe, matches, and some provisions. Then I began my trek to McMurray, staying on the river to avoid getting lost.

The snow was deep and powdery, and, even with snowshoes, I sank with every step. After travelling a quarter-mile or so, it became apparent that my decision was not a wise one: I would never make McMurray. So I retraced my steps to the aircraft and made another attempt at radio contact, still without result. I laid the wing covers out in the cabin and stretched my bedroll on top of them. After cutting some wood on the riverbank, I lit a fire outside the aircraft in the tin stove then made some tea and soup for supper. The weather was clear and very cold, so I used the blow pot to heat up the cabin. Even though I had a double-layered sleeping bag, I could not keep warm.

On Saturday, 1 February, the weather was completely socked in with ice crystal fog right down to the river. The ceiling and visibility were zero, and the temperature was right off the gauge. It was still very cold when I got up at 10:30 and lit the blow pot to make some tea. I could not get my feet warm—something had to be done. So I set up the tin stove in the cabin, punching a hole in the Perspex of the astrodome for the stovepipe. This helped a lot. I climbed the bank to gather firewood; but after a few strokes, the axe handle broke, limiting my wood source to the smaller branches that I could break by jumping on them. It also meant that I had to cover a larger area just to gather an adequate supply. At 12:15, the temperature was -45 (F). While there was about a seven-day food supply in the ration kit, I kept my meals light as a conservation measure. My rations consisted of tea, hardtack, dry soup, and some canned Spam. In an effort to get warm, I went to bed at 17:00. I had been told that the best way to conserve body heat in a sleeping bag was to remove all of one's clothing and climb in. So I tried it. The author of that theory must have conceived it in Florida. In the end, I put on everything I had: heavy underwear, thick wool pants and shirt, heavy winter parka, and even flying boots. Both wing covers had been serving as my mattress; now I pulled one over me as further insulation.

By Sunday, 2 February, the clouds were starting to break up—yet the temperature remained off the gauge. I got up at 2 AM and lit a fire in the stove. Condensation was dripping from the cabin roof. I woke up again at 8:10 and found that my cheek had frozen to my sleeping bag where my breath had condensed. After carefully thawing the ice with my hand, I got up and made some tea and tomato soup for breakfast. Then I climbed the bank to gather more firewood. Happily, the cloud cover was steadily dissipating. At 16:00, I made tea again to wash down the Spam I fried for supper. A half-hour later, I heard a

Associated Airways' unique clipped-wing Avro Anson V with Tommy Fox at the controls, moments after its arrival in Edmonton on 5 February 1947 from Fort McMurray, Alberta. *T. Fox*

sound outside and opened the door to see the Taylorcraft CF-DBY passing overhead. There followed much waving of arms on the ground and finally a big sigh of relief after a waggle of wings overhead. Maury Danes landed and taxied alongside. He joined me for a cup of tea. Then we doused the fire and took off for McMurray at 16:45. We landed at the airport at 17:15 and wasted no time getting to the hotel in town, where I had a bath, shave, and supper. Then I sent a wire to Associated at Edmonton requesting an Anson to bring a drum of gas and a jeep heater. After receiving the wire, Associated advised my wife, Clara, that I had been missing but was now safe in Fort McMurray.

On Monday, 3 February, the McMurray temperature was -40 (F). The Anson, piloted by Doug Ireland, arrived at 12:45, and we departed for 'EKD without delay. Immediately on arriving, we put the jeep heater on the port engine while we pumped the drum of gas into the starboard tank. When this was done, we switched the heater over to the starboard engine, removed the port engine cover, pulled through the engine by hand, and then started it. It took considerable time to thaw out the starboard engine; but, when heated sufficiently, it too was pulled through and started. After both engines were brought up to temperature, we applied power. By using rudder and elevator to "bump" the ship, I broke it free and moved forward slightly. We decided that Doug would take off and make a long run before lifting off in order to create tracks that 'EKD could follow. Both takeoffs were uneventful, and we flew to McMurray, landing at 17:15. Associated at Edmonton was advised of our arrival time by telegram.

It is of interest to note that on 3 February 1947, Snag, Yukon, recorded an all-time North American low temperature of -81 (F). The Edmonton weather office estimated temperatures of -70 (F) at the site of my forced landing.

CONCLUSION AND AFTERMATH

On Wednesday, 5 February, the weather at McMurray and Edmonton was clear. We warmed up the ships and prepared them for takeoff. As Doug would be cruising faster than 'EKD and arriving at Edmonton first, he was to tell the people there to make room in the hangar and open the doors as soon as I landed so that 'EKD could taxi straight in without stopping. This was to avoid parking outside and attracting a

Map of northern Alberta showing the route flown by CF-EKD. The legs from Chipewyan Lake to Fort McMurray and from there to Edmonton were flown with the aeroplane's drastically reduced span. *B. McLellan*

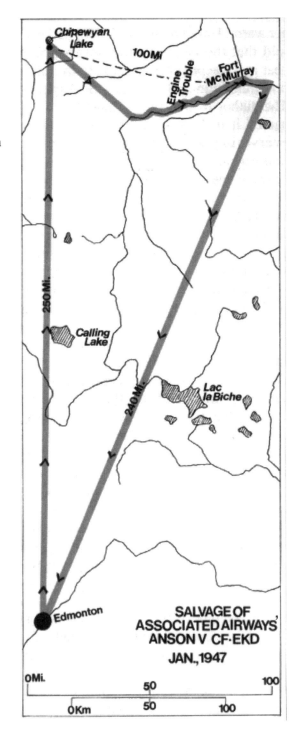

Chipewyan Lake

100Mi.

Engine Trouble

Fort McMurray

250 Mi.

Calling Lake

240 Mi.

Lac la Biche

Edmonton

SALVAGE OF
ASSOCIATED AIRWAYS'
ANSON V CF-EKD
JAN., 1947

0Mi.   50   100

0Km   50   100

128

crowd. I took off first in 'EKD and headed for home. I encountered moderate turbulence en route and the effects of the shortened wings and reduced aileron surface became apparent. It took full travel of the controls to keep the ship right side up! The turbulence subsided before I reached Edmonton, and I encountered no trouble landing.

'EKD had performed much better on this leg of the trip. With skis off and landing gear retracted, I could easily maintain an air speed of 100 mph plus. Since the radio had decided to work, I had normal contact with McMurray and Edmonton. After landing at Edmonton (flight time: two hours and 30 minutes) I taxied to Associated's hangar; the doors opened as I approached, and I went inside. As soon as I shut down the engines, the hangar doors were closed. We were home and safely out of sight—or so we thought.

I climbed out of the ship to congratulations and handshakes from the Associated gang, then headed to the office. I had no sooner arrived when Jock Currie, the Department of Transport superintendent of engineering and maintenance, burst in wanting to know what the "God-damned contraption" was that had just flown over his brand new $10,000 house. To say that he was unhappy would be an understatement. He demanded that my pilot's and aircraft engineer's licences be surrendered to him on the spot. Although we showed him the aircraft logbooks, particularly the entries that certified the aircraft as "not airworthy" and "not fit for flight," he would not be dissuaded from taking my licences. He stated flatly that licences were only issued to responsible people and that no responsible person would fly a "God-damned contraption" like that.

The next day, I received a phone call from Ken Saunders, district superintendent of air regulations, requesting that I come to his office for a meeting. There he expressed his opinion of my unapproved modifications and advised me that he was sending my licences to Ottawa headquarters with a recommendation that they be suspended. Happily, about a week later, Ottawa returned the licences with neither suspension nor comment.

What finally happened to 'EKD? In 1947, War Assets was selling Anson Vs in three ways: with logbooks for both engines and airframe, with logbooks for engines only, or with a logbook for just the airframe. With only one set of books, the cost was halved. Since 'EKD's engines had very little time on them, we bought a low-time airframe and installed them. 'EKD was also fitted with a cargo door, so we used that section of the fuselage in the new ship. In this way we ended up with a large number of legitimately airworthy parts, instruments, and accessories. In short, 'EKD, with the exception of the main wing section, ended up in our stockroom as a valuable asset.

Hudson Bay on Skis and Floats

Year-round Operations over a Vast and Forbidding Territory

GEOFFREY (JEFF) WYBORN

Jeff Wyborn was one of those quietly unassuming yet highly capable people whose greatest satisfaction is in knowing they did their job well. And his job was not an easy one. For most of his 35 years with Austin Airways he serviced the remote, mostly Native communities on the shores of Hudson Bay, flying a Norseman. He accumulated nearly 10,000 flying hours on Bob Noorduyn's masterpiece.

Jeff pioneered operations into locations as far north as Winisk and Fort Severn on the western shore of Hudson Bay and, on the east side, to Fort Smith (now Inukjuak), Quebec, and beyond. Conditions over this route of some 18,000 kilometres were harsh and unpredictable, and the terrain was vast and barren with few landmarks. From each coastal stop, lengthy trips inland were always required; winter and summer, the demands on pilots were daunting.

Because Jeff and his fellow Austin pilots such as George Charity, Hal McCracken, and "Rusty" Blakey provided so many vital services and were the region's most tangible link with the outside world (in the days before satellite television), they became local legends.

Jeff Wyborn was raised in Port Arthur, Ontario (now a part of Thunder Bay), and learned to fly with the old Thunder Bay Flying Club, earning the money for his flying lessons by selling newspapers and running a 5 AM milk route. He was 17 when he soloed in 1937.

He obtained his air engineer's licence while building Hawker

Hurricanes with the Canadian Car and Foundry during the early war years. Like so many pilots of his day and of the previous era, he was able to maintain as well as fly aircraft. Such dual qualifications could prove useful if a pilot was forced down miles from help. They also made it easier to find work. During World War II Jeff served as a maintenance crew chief on Tiger Moths with No. 2 Elementary Flying Training School in Fort William, Ontario. He worked for Canadian Airways and the Ontario Provincial Air Service—flying Gipsy Moths and Buhls—before joining Severn Enterprises, where he begins his story.

Jeff Wyborn retired from commercial flying in 1985 and shortly thereafter addressed the Toronto Chapter of the Canadian Aviation Historical Society. His talk, in much expanded form, became the substance of the following account, which appeared in the Spring 1988 CAHS *Journal*. More informative than anecdotal, Jeff's story is remarkably sparing in its use of the personal pronoun "I"—indicative of his natural modesty. His understatement, not unlike that of RAF pilots in World War II, reflects his calm, matter-of-fact acceptance of the never-ending challenges of flying in the far north.

Jeff Wyborn died on 24 November 1993 at South Porcupine, Ontario, where he had made his home for over three decades.

It was mid-afternoon, 26 September 1950, and I was about to depart on my last flight for Severn Enterprises before joining Austin Airways. I had been three years with Severn as a pilot/engineer flying charters and delivering supplies and trade goods to trading posts, with a back haul of furs. In the summer, we carried similar cargoes, on floats, to the posts and to Indian fishing camps, where we collected back hauls of sturgeon to be dropped off at Sioux Lookout or Hudson, Ontario, via Pickle Lake. When flying activity was slow I performed maintenance duties, assisting Bill Williams, the base engineer. We flew a J-3 Cub, an Aeronca Sedan, and I occasionally co-piloted the Travelair 6000, CF-AEJ. Sometimes we would rent a Northwest Industries-built Bellanca Skyrocket, CF-DCE, from Don Hurd, a prominent mining executive.

I was at Pickle Lake with a load of sturgeon destined for Sioux Lookout, enjoying a bowl of soup and a sandwich at our "home away from home," Mrs. Koval's hotel. Another Severn employee refuelled my aircraft, the Aeronca Sedan. I took off, and about 30 minutes out of Pickle Lake, the engine sputtered and quit. I dead-sticked a landing on Miniss Lake, where a strong northeast wind drifted the aircraft into a rocky bay, providing me with a reasonably safe mooring. The weather turned sour; the ceiling lowered. Soon it began to rain, and the fog rolled in. I was cold and miserable for three days until Jimmy Kirk, flying for the Ontario Provincial Air Service, spotted my aircraft and landed to see if I needed help. We flew back to Pickle Lake, where I collected some gas, and then he returned me to my machine. Gassing up, I completed the trip to Sioux Lookout without further problems.

Why had I run out of gas when the aircraft had been refuelled only a half-hour earlier? The man on the pump had neglected to put the gas caps back on, and the airflow had sucked all the fuel out of the open spouts. Normally I checked to make sure that gas and oil caps were secure—except this time.

AUSTIN AIRWAYS, A PIONEER NORTHERN OPERATOR
I began flying for Austin Airways on 4 October, operating initially from their Sudbury, Ontario, base with three other pilots, Chuck Austin, "Rusty" Blakey, and Hal McCracken. A month later I moved to what was to become my permanent base at South Porcupine, where my wife and two children joined me.

We flew the Noorduyn Norseman, designed by Robert Noorduyn specifically for Canadian operations and built in Montreal. The Norseman was well suited for the type of flying we did, operating over the rugged inland and coastal areas of Hudson and James Bays. It was structurally very sound and performed well—provided there was no excessive overloading. The Norseman was a challenge to fly, more so than the later Beaver and Otter aircraft, which were relatively forgiving. It required power when landing with a load on floats or skis, to decrease the angle of approach. As an illustration of how docile the Otter was, Jimmy Bell once needed someone to fly in a first seasonal load of tourists from Moosonee, Ontario, and I happened to be available. I hadn't flown an Otter for 10 years, and told him so. He asked me what I was flying at the time, and when I mentioned the Norseman his only comment was that the Otter should be no trouble at all—go ahead and take it up. Yet I found it a pleasure flying all three types.

Austin maintained two Norseman aircraft on the east coast of Hudson Bay and one on the west coast to service the annual goose hunt from Moosonee as well as handle general coastal flying, and the inland freighting for the Indians. In a pinch, we could count on Gold Belt Air Services with pilot George Pauli and engineer Rolly Denamie flying their Norseman out of Rouyn-Noranda to help us out at Old Factory and Eastmain.

Float operations usually began as soon as we were advised by Hudson's Bay Company radio that the rivers at Albany and Attawapiskat on the west coast of James Bay and those at Rupert House, Eastmain, Old Factory, and Fort George on the east coast were free of ice. Two Norseman aircraft would depart from Sudbury, stop at South Porcupine, and then proceed to Moose Factory. The South Porcupine pilots included George Charity, Jay Pipe, and me. At this time, in the early 1950s, each pilot was also a licensed air engineer (AME), expected to assist the crewman with maintenance duties and responsible for signing off the aircraft as airworthy.

Moose Factory became our base of operations for trips up both sides of the bay. We were provided with comfortable accommodation and our meals at the Hudson's Bay Staff House. Pilots and crewmen, we represented Austin Airways as public relations officers, dispatchers, and freight-handlers—as well as maintaining our aircraft. Our principal customers were the Departments of Indian Affairs, Health and Welfare, and Transport (radio and weather stations), and the Post Office. We also serviced the Hudson's Bay Company, the free traders, the Anglican and Roman Catholic missions, and Native people from the many settlements in the region.

Trips northbound were either on a charter or a passenger-and-poundage basis. Under the latter arrangement we would depart for points north as soon as a load had been accumulated, weather permitting. In many cases the bulk of the load would be mail, with possibly two to four passengers. We were always very careful to make sure that all first-class mail was aboard. Hudson's Bay posts on the west coast of James Bay and Hudson Bay, such as Albany, Attawapiskat, Weenusk (renamed Winisk), and Fort Severn, were serviced by Austin from Moose Factory. Those on the east coast: Rupert

Two popular modes of northern transport meet at the Moose Factory
Hudson's Bay Company post in the early 1950s: Jeff Wyborn's
Norseman and one of Austin's Bombardier snowmobiles. *J. Wyborn*

House, Nemaska, Eastmain, Factory River, Fort George, Great Whale,
Richmond Gulf, Port Harrison, Povungnituk (Pov), Cape Smith, and,
on occasion, the HB post on the Belcher Islands, were also handled
from Moose Factory.

By the late 1950s coastal flying on both sides of Hudson Bay had
become a daily, scheduled service. We used single-engine aircraft initially but eventually replaced these with twins, such as the DC-3. These
operated on wheel-skis from landing strips marked out on river and
sea ice for many years, until year-round landing strips became operational at most settlements in the North. Tribute must be paid to our
predecessors, such as Canadian Airways, who pioneered charter flying
into the region during the '20s, '30s, and '40s with earlier equipment
and under even more primitive conditions.

OUR DIVERSE CARGOES, CUSTOMERS, AND PASSENGERS

Our cargoes could include bulky items such as bags and still more
bags of eagerly awaited mail, bales of fur (just as bulky), and even
containers of live beaver for restocking trapped-out or depleted areas.
When we had heavy equipment to carry, such as diamond drilling

motors, pumps, rods, casings, and 45-gallon drums of fuel oil and gasoline, we would always lay a false floor of one-half or five-eighths-inch plywood to protect the actual floor of the aircraft and the underlying structure. All loads had to be properly secured; any movement could endanger the aircraft and those on board. External loads, such as freight canoes, lumber, and long rods, were carried regularly. The canoes were always carried stern first, on either the port or starboard side of the aircraft, their load factor being based upon drag plus the actual weight. The internal load would be cut in half to maintain the legal limit.

There could be problems. On occasion, Native people being flown to their trapping grounds might become sick. Or their dogs would begin to fight—big sled dogs—the din could be heard above the engine noise! Their crashing about the cabin, especially on a long trip, would make it necessary to compensate by continuous playing with the elevator trim, nose-up or nose-down, to counteract the rapid weight shifts.

On numerous occasions, messages through the Hudson's Bay Company network in Morse code would request an aircraft for a medical evacuation. Some of these flights were to the eastern Arctic, to places as remote as Port Harrison, Povungnituk, or Cape Smith. Many southbound passengers were Indian or Eskimo patients to be admitted to the Moose Factory Indian Hospital; TB in various stages was extremely prevalent among Native peoples. Coughing seemed to be continuous; every child appeared to have a runny nose. It always amazed me that, in spite of our close associations with infected patients during these many flights, none of the pilots or engineers contracted the disease.

Mining was another source of work, and in the 1960s it increased substantially on the Belcher Islands and along the Povungnituk Range, as far as the inland area at Esker Lake and the famous Chubb Crater. Flying over these regions, one could always spot the red tents used by Murray Watts, a mining-exploration company doing extensive work in the Mosquito Bay and Povungnituk areas. Farther south there was prospecting around Sakami and Old Factory Lakes, and about 150 miles inland, a company was searching for diamonds along the Eastmain River. We flew in hundreds of tons of freight to these various settlements on both coasts during the winter and summer months. This was well before the building of local landing strips for year-round access—we operated only on skis or floats.

Under the sponsorship of the Department of Indian Affairs, the Indians set up sturgeon fishing camps at the mouths of the Nottaway, Broadback, Pontax, and Harricanaw Rivers to provide seasonal employment. These large bottom-feeding fish are equipped with mouths that resemble nothing so much as the top of a mukluk, and are caught in nets or with night lines fitted with large baited hooks that rest on the bottom. They are a source of caviar, and their flesh is considered a delicacy. They are parboiled to rid them of excess fat, and then fried as steaks. Smoked sturgeon is also superb in taste and texture, with no ill smell. In either case they are gourmet food. When caught they were kept alive in submerged log cribs or tied on long cords in shaded areas of the stream until the aircraft had landed—then they were slaughtered. Flown to Moosonee twice a week, they were shipped to points south by the Ontario Northland Railway. The roe, or caviar, was removed, and shipped with the fish, usually to New York. The Native fishermen retained the heads, which they regarded as the best part of the fish, an even greater delicacy.

During September and as far into October as the weather permitted, we flew supplies to Native trappers and their families for the upcoming season: food, equipment, dogs, and canoes. During that period as many as 50 such trips might be completed from Fort George alone, the largest settlement on James Bay north of Moose Factory. These inland trips ranged anywhere from 60 to 300 miles, the longest being to Opimiskow Lake. On one such trip we followed a map hand-drawn by the Indians, with written notations added by the various pilots who had used it. While we flew inland from other settlements, Fort George always required the most trips.

The previously mentioned goose hunt was an Ontario Northland Railway operation along the shores of the Harricanaw River, which we serviced from their "end-of-steel" at Moosonee. Another lodge at the mouth of the Nottaway River, owned by C. S. McLean, was also visited by Austin aircraft, as was one at the mouth of the Albany River operated by Len Hughes. When scheduling our trips inland, we always arranged to handle the more northerly ones first because these lakes and rivers froze over at an earlier date. One of the free traders, George Papp on Richmond Gulf, required trips of 100 to 150 miles.

Bill Anderson was another well-known free trader and goose camp operator who chartered inland trips from Fort Albany.

In the early 1950s dental, medical, and X-ray teams began operating out of Moose Factory Indian Hospital under the supervision of Dr. Bert Harper. The program was initiated because of the high incidence of TB among the Natives. Local Peterhead boats supplied by the RCMP, the HBC, and others transported the medical teams. These boats, presumably first built at Peterhead in Scotland, were also constructed on Canada's east coast and at Moosonee and Moose Factory. The teams travelled deep into the eastern Arctic to Port Harrison, Povungnituk, and to every settlement on James and Hudson Bays under their jurisdiction. They relied upon our Norseman aircraft to resupply them with food and medical needs. The residents of these settlements were X-rayed and given medical and dental examinations. We flew in X-ray plates and brought the exposed ones out to the hospital at Moose Factory. On a few occasions we flew teams and their equipment from Rupert House to Nemaska, 80 miles inland. Over the years the Indians and the Eskimos have been largely cured of TB, and this must be credited to the efforts of these medical teams, led by Dr. Harper, his successor, Dr. Macon, and their base hospitals.

OUR HUDSON BAY AND INLAND ROUTES

A season's breakdown of our inland flying from only the busier settlements might be as follows: Fort George, 50 trips; Old Factory, 40 trips; Eastmain, 30 to 35 trips; Rupert House, 20 to 25 trips. When a Norseman arrived at Fort Severn, one of the smaller locations, and off-loaded passengers, it was positioned to begin the inland freighting for the region. This would involve three or four trips and, if the weather was bad, might take five or six days to complete.

Our next stop southbound would be at Winisk, where the Indian people would require about four inland trips, ranging from 100 to 140 miles. These completed, we would depart Winisk to land at Hawley Lake, where we might take on board a few members of northern families such as the Chockmolons. Our next stop would be at Attawapiskat to refuel and board more passengers, often patients and their luggage—providing we did not exceed our all-up weight. The last two stations to have their inland freighting completed because of their more southerly location were Fort Albany and Attawapiskat on

Local Natives cluster around Jeff Wyborn's Norseman at the Ghost River Inuit settlement in the winter of 1954. The women had abandoned skins, except for fur trim, replacing them with colourful coats and materials purchased from the Hudson's Bay Company.
J. Wyborn

the west coast of James Bay. The Hudson's Bay Company sponsored the flying requirements of those Natives who trapped for the company. Norsemans carried schoolchildren from the Bishop Horden School at Moose Factory back home to Mistassiny, a distance of 300 miles into the Quebec interior. A few years later, Austin's Cansos would take over this job. Austin Airways also transported other schoolchildren from the various settlements to their homes in the James Bay and Hudson Bay areas.

All flights to the eastern Arctic settlements would be completed by the end of September, we hoped. On longer flights, to "Pov" and to Port Harrison, we overnighted at the Department of Transport (DoT) weather stations and at the HBC residence. Art and Mrs. Bambrick and Mr. Shackleton were the DoT marine and air radio officers, while Rube Ploughman and Ralph Knight were HBC managers, all residents of Harrison. At Great Whale we usually stayed with Mr. and Mrs. Fred Woodrow, who operated the DoT's weather and radio station. One of the attractions at their place was a pet Arctic white owl, alleged to be friendly—but it sure didn't look it. Many other people also generously fed and housed those of us who did the flying around Hudson Bay, including the Department of Health and Welfare Nursing Stations, Hudson's Bay managers and their wives, the DoT air and marine radio stations, the Anglican and Roman Catholic missions, the free traders (Bill Anderson at Albany and Mr. and Mrs. George Papp at Richmond Gulf), Mrs. Maude Watt of Rupert House, and the locals at Moose Factory and Moosonee.

I did the bulk of my work near James Bay and points north. The flying activity from South Porcupine mainly involved Beaver and Cessna 180 aircraft flown by pilots with less experience in the North. In spring, summer, and fall these aircraft were kept busy during most of the daylight hours, seven days a week, weather permitting. At times it would be necessary to call in other Austin Beaver and Norseman aircraft to help out.

As for our fuel arrangements, we kept our own large gas caches at all Hudson's Bay Company posts, and maintained a constantly updated inventory of the quantity at each location. We did not want to be caught short. Hudson's Bay Company supply ships such as the *Fort Charles* and the Peterhead boat *Agamaski* from Moosonee brought in aviation gas and oil to many of the settlements in the James Bay and Hudson Bay region. Ocean-going vessels from the east coast of Canada supplied the eastern Arctic settlements. Roman Catholic mission vessels also carried some of the fuel. Much of it would be used during the fall inland supply, the coastal flying, and during the winter fur pick-up.

FLIGHTS COULD BE MEMORABLE—FOR MANY REASONS

When the season's flying was finally completed, we immediately began the trip south to our main bases. Operations near the end of the season were always challenging and unpredictable. I once had a load of 1,400 pounds of supplies for a Native trapper and his family who were located on an unnamed lake about 100 miles from Moose Factory, toward Hearst. When I arrived over the lake, on floats, I found that it was frozen—there was no possibility of landing without mishap. I had no choice but to continue on to South Porcupine, where we contacted the RCAF. They agreed to drop the supplies by parachute, and did so successfully. I believe they used a DC-3 from

Trenton, Ontario. The air force ran an efficient, well-organized program. From then on this lake was known as Parachute Lake.

One of our more adventurous trips was to the Hudson's Bay post at an Eskimo settlement on the Belcher Islands, a flight of 100 miles over open sea. The Belchers are a very large and desolate mass of rocks that has withstood the scouring of continental glaciation that once covered the area. The deep grooving that resulted is a prominent feature, actually more typical of the Arctic geography farther north. This trip was about as far as I would care to fly over open water in a single-engine machine like the Norseman, and was only undertaken under the best of weather conditions. Reports on weather were broadcast by the DoT marine and air radio stations at Great Whale and Port Harrison, and from a small DoT station at Moosonee. The HB managers transmitted additional information on weather and surface conditions at the various settlements during their scheduled broadcast times, and on special request.

But the weather was always a gamble. On one occasion we departed Port Harrison for Great Whale and points south. Since there was a high overcast and good visibility, we observed our usual practice of following the shoreline. Although doing so added to our mileage and flying time, it improved the safety factor. The distance to Great Whale was approximately 275 miles, and, as we progressed, we checked our position regularly. At about 85 miles, the weather began to deteriorate; the wind from west-northwest brought in low cloud and fog from the bay, and we were forced to land in Nastapoka Sound. Whenever the fog would lift sufficiently, I would take off, working my way south by following the coast, always with an eye on the weather, landing whenever fog threatened to close in. We continued taxiing, trying to locate a suitable beach on one of the coastal islands where we could bring the aircraft ashore.

At one point, taxiing in heavy fog, I inadvertently ran through an opening between two of the islands and found myself in the open sea of Hudson Bay. Swells were running from five to seven feet, dangerously high for a floatplane. But I managed to keep the aircraft afloat until the fog lifted enough for me to take off—crosswind—along the crest of a swell. I landed back at Nastapoka Sound, and we taxied for a further two or three miles with decreasing visibility before we finally found a suitable beach on Anderson Island. Darkness immediately closed around us. Since this was tidal water, subject to changes in level, we collected rocks to secure our tiedown ropes. Fortunately, we had a light load and no passengers. My crewman, Tom Nelson, and I spent a cold night pondering the uncertainty of life in general, and of Hudson Bay weather in particular. The next day we were able to reach Great Whale—where we stayed put!

ROUTINE HAZARDS AND DISCOMFORTS

Some of the settlements were a challenge to get into and out of: Weenusk (later Winisk), on the river of the same name flowing into Hudson Bay from the west, had about the worst takeoff and landing area. We had to use a fast-flowing stretch of the river, about a half-mile in length, between two sets of rapids. Bringing the aircraft around into wind, just before applying power for takeoff, a pilot's judgement had to be right on. In the process of turning, a lot of distance could be lost to the fast current. And when adverse winds combined with the swift water, aircraft had been known to run the lower rapids unintentionally. Damaged floats were the usual result.

Occasionally in late autumn, we would receive a heavy overnight

Jeff Wyborn's Noorduyn Norseman, CF-GMM, during winter deliveries to a settlement 120 miles inland from Fort Severn on the west coast of Hudson Bay. *J Wyborn*

fall of wet snow that collected on the upper surfaces of the aircraft. This added weight would force the floats low into the water, sometimes half-submerging them. Whether the aircraft was moored at a buoy or tied to a pier, we would be down at all hours of the night, cautiously removing the snow. If the buildup was great, we did not dare to climb aboard lest our added weight cause a float to sink and capsize the aircraft.

Mooring in the tidal water at the mouth of a river where a settlement was located could be tricky. If there was much wind the aircraft would weathercock into the wind, and you could find yourself drifting sideways—rapidly—with the tide or current. Manoeuvering became very difficult indeed. In a tidal area it was necessary to come to a stop and then observe a fixed point on shore to establish the exact direction of your drift. We would bring the aircraft ashore, well above the high tide mark, and tie it down for the night before setting to work making camp.

The mosquito posed a different sort of problem. These persistent little insects found Port Harrison and Povungnituk, surrounded as they were by barren, rolling hills, to be a haven. Whenever we flew there in summer we would tie our aircraft to the dock, disembark our passengers, and unload our freight—by which time we would be able to hear the welcoming party of Eskimos. While we could not see them behind a rise in the ground, their position was apparent from the black cloud of mosquitoes hovering above their heads. Yet people native to the area never seemed to be bothered by them, which was not the case with white transients like us. The lowlands of the west coast of Hudson and James Bays were ideal breeding grounds for mosquitoes and black flies. They plagued us as we unloaded the aircraft and closed the doors as quickly as possible, but we still managed to provide free transportation for any number of these pests. We always hoped—and prayed—that a wind would come up, just long enough for us to complete our refuelling or any necessary work on the aircraft.

Winter flying introduced a new set of hazards. We often encountered slush on certain lakes; an aircraft might become mired in a combination of deep snow, slush, and water (layered, in that order) to a depth of two to two-and-a-half feet. This invariably meant hours of hard slogging, jacking up the aircraft and cutting logs in the bush, which were laboriously dragged back to the machine and worked under the skis. In the process you became soaking wet—from perspiration and the water you were slogging through—in temperatures that could drop to -15 or even -40 (F). Then it might be necessary to spend the night preparing a path immediately in front of the machine, followed by a chilling wait for snow conditions to improve sufficiently to attempt a takeoff run. Situations like this made the glamour of bush flying more myth than fact.

We regularly faced winter temperatures of -40 (F), and conditions that dictated extra chores and precautions. A cover had to be placed over the engine and sleeve-type coverings pulled over the wings; the former kept the weather out of the engine and the latter prevented a buildup of frost on the wing surfaces. Even a thin, seemingly insignificant film of frost on the wing could drastically reduce its lifting ability and result in a difficult takeoff or even a serious mishap. Planks or logs had to be inserted between the skis and the snow; if this were not done, the skis would freeze in place, and would have to be broken loose and then cleaned thoroughly. Before takeoff, plumbers' blow pots filled with naphtha gas were used to heat the engine. We would light them, and when the blow pot's generator became heated, we opened the valve, producing a hot blue flame. The blow pots had to be pumped regularly to keep up the pressure. We would sit beneath the engine cover with two or three of these roaring away and an extinguisher handy—too many aircraft had been lost to fire during this procedure. Periods of such pre-heating varied from 25 to 45 minutes, depending on the outside air temperature and the wind-chill factor. In extremely cold spells we removed the oil and battery, and carried them indoors.

WINTER OPERATIONS

As soon as the ice in the Sudbury and South Porcupine areas, where Austin bases were located, was thick enough, our aircraft were converted for ski operations. All of the managers of the Hudson's Bay posts on James Bay kept Moose Factory, South Porcupine, and Sudbury advised as to the thickness and general condition of the river ice where our takeoff and landing areas were subject to the effects of fast currents and tides. When we received the green light, two or three aircraft would fly to Moose Factory to commence winter ski operations, usually on one of the first days of January.

At Moosonee, we parked our aircraft, tied them down, and then set to work. We would contact Mrs. Louttit, the postmistress there, and be informed that she had a post office full of mail for us to deliver, to be added to the other cargo that had accumulated. We set out up both coasts with Christmas mail, freight, and passengers. Southbound, we returned with more mail, patients for the hospital, and their luggage. Once the rush of coastal activity had settled down, we began flying inland to the Indian camps last seen in the early fall.

We began a series of visits to Richmond Gulf, Great Whale, Fort George, Factory River, Eastmain, and Rupert House on a pre-arranged schedule, set up by the Hudson's Bay managers and the Indian trappers before our departure in the fall. On every flight inland we carried a full load of grub: 100-pound bags of flour and sugar, cases of lard and canned milk, as well as other staples and some "goodies." We would land at the various camps and unload the supplies, then reload with beaver, mink, fox, wolf, and other furs that had been trapped, as well as the occasional ailing family member who had to be flown out to Moose Factory for treatment.

During our circuits of the Hudson's Bay posts we would often meet up with their company pilots and engineers: Will Kennedy, Art Atkinson, Slim Belcher, Al Snyder, Fred Bradford, and Les Templeman, among others. Their aircraft were the familiar silver Norseman CF-BHT and 'BSL and their DHC Beaver CF-FHU.

One January, I departed Moose Factory in my Norseman with mail and freight for Winisk and Fort Severn, stopping at Attawapiskat to refuel for the final 250 miles to Severn. Since gasoline was very expensive north of Attawapiskat, we used no more than was necessary. We had left Moose Factory on a clear, cold morning, with the temperatures at -25 (F). By the time we reached the Winisk area it had dropped to -50 (F). I noticed that the cylinder head and oil temperature were at the lower end of the operating range, while the engine oil pressure was much above normal. The extreme temperature had caused the oil to thicken, hampering circulation. Since I had no place to land, I pressed on toward Winisk, which was about 15 or 20 miles ahead. Suddenly all hell broke loose: oil leaking through the seals in the propeller splattered the windshield with a tarlike layer. With no vision ahead, I was able to scrape enough congealed oil from a side window to reach Winisk, where I managed a rather tricky landing on the 1,200-foot strip of packed snow marked out between the two pre-viously mentioned rapids, now frozen, but just as dangerous. When the oil seals gave, I was reminded of the truth of that old axiom: A chain is only as strong as its weakest link.

On another winter occasion, as I was flying east of Moose Factory over the James Bay lowlands, I decided that it was about time to change from left to right wing fuel tanks for the remainder of the flight home. As I turned the fuel selector from one tank to the other, part of the system broke—effectively shutting off the fuel supply to the engine, which responded by quitting. I trimmed the Norseman to a nose-down attitude to retain control more easily while I ripped off a side panel. With my pliers, I succeeded in turning the selector part-way on. Although I had not known which way to turn it, I had guessed right. As luck would have it, I twisted it just far enough to allow fuel to reach the windmilling engine, which caught barely in time to keep us out of the trees. With the selector not in the fully open position, I had to operate the engine at a higher boost and lower rpm with full rich mixture to keep it running until we got back to Moose Factory.

"Improving" our winter landing areas was a standard practice of the Hudson's Bay Company. On the day previous to a planned flight, they would hire the local people to tramp down a strip on which we could safely land. Any drifts of hard, wind-blown snow would be cut away and levelled, with small trees placed to mark the perimeter.

Regarding our landing gear, the type of aircraft ski that proved to be best suited for arctic use was a combination of the M & C (Mayson and Campbell) pedestal, with a ski made by Elliott Brothers. The ski was constructed of laminations of hardwood (ash) sheathed with brass, while the pedestal contained a tough rubber bladder on which the undercarriage axle rested. With the shock-absorbing ability of the inflated bladder added to that of the Norseman's normal oleo leg, the undercarriage was capable of absorbing the sort of punishment that came with arctic flying. Locals were not always at hand to flatten run-ways for us, and surface conditions could be severe, with ribs of hard, dry snow. Those M & C pedestals and rugged Eillott skis prevented much structural damage to our aircraft.

Another good ski installation for arctic use was that manufactured by Northwest Industries of Edmonton. These wheel-skis, called *Roll-ons*, were satisfactory for operating from an airport, and for the hard-packed snow conditions prevalent in the Arctic. But they did not have enough surface for some parts of the North—they would bury themselves in deep, soft snow. Previous to the M & C pedestal, which came to be used on all types of bush aircraft, pedestals had been made of solid, laminated hardwood, of steel tubing (with and without springing), or of formed and welded metal in great variety.

WHEN ENGINE STARTS WERE A CHALLENGE

On skis, whenever we experienced a dead battery or starter trouble, we knew it would be difficult to get the engine running; but we always managed to do so by the risky process of hand-swinging the propeller. Normally, I would be at the controls with my crewman swinging the prop; but if I had put the crewman at the controls my signals to him would be: "Gas on—switches off (master and ignition)," followed by four or five priming shots during the propeller pull-throughs. Then, making sure the propeller was in the correct position for hand-swinging, I'd yell out: "Gas on—throttle set—contact! (switches on)." Then I'd swing the prop. Sometimes it started, sometimes it didn't. In the latter case, it was the same procedure all over again.

When I've had no crewman with me, I've managed the hand-swinging business by myself. With a Beaver, I've hand-swung the prop from the front. (A Beaver on skis does not sit as high as a Norseman.) Fortunately, it was not something that had to be done often. The practice would only be used to get the aircraft back to a settlement for a fresh battery and/or a starter and/or a generator replacement.

Why didn't I make use of the hand-cranked inertia starter? We had removed these from our aircraft because they had not proven very useful. If you are drifting on a river or with the tide, cranking the starter and then hoping the engine would start was a chancy business.

The process was more complicated when hand-starting the Norseman on floats fitted with a three-bladed propeller. We began by tying the aircraft parallel (more or less) to the shore or dock, with a mooring rope at the rear float strut or rear bollard, and the nose pointed slightly out. The crewman dropped a slip-knotted rope over the top of the number-one blade, which was positioned in front of number-one cylinder for firing, just a few degrees past top dead centre. As he pulled the rope to swing the propeller, I would be set to start; gas on, throttle set, switches on (contact)—and away she would go—maybe! When the motor caught, the crewman released the mooring rope and hopped aboard. For all hand-swung starts, I set the propeller blades in fine pitch. We never grasped the three-bladed prop on the Norseman—it was relatively high, and the next blade came around too fast, once it caught. But I have hand-swung the prop of a Beaver in summer. Standing on the float, behind the prop with the crewman at the controls, I've pulled it down. Once it caught, I would climb in, take over the controls, and take off.

The major cause of engine fires while starting has always been over-priming, whether in summer or winter. When the pilot has made a few attempts to start his engine but does not succeed, there is a possibility of a carburettor fire. The advantage of a direct-drive starter is that the pilot or engineer can keep the starter in operation, turning the engine over to suck the flames into it and possibly preventing a serious fire. Like most other pilots and engineers, I have had several such annoying fires.

SOME GENERAL COMMENTS ON THE ARCTIC I KNEW

In addition to the aeroplane, people in the North also relied on the canoe for transportation in the summer. The canoe was the traditional watercraft of the Indian, and over the years the design has been perfected. The Hudson's Bay Company had established a small commercial factory at Rupert House, employing Native craftsmen, who turned out canoes ranging from tiny 12-footers to the very practical square-sterned 22-foot freighters. These were heavy, well constructed, and very stable—strong enough to withstand the rough and treacherous seas off the coasts of James Bay. While the canoes could be paddled, as in past centuries, they were usually powered by outboard motors.

The Eskimo watercraft was the seemingly fragile kayak, built with a minimal framework of wood covered by skin. At places like Port Harrison, Quebec, we occasionally watched Eskimos indulging in impromptu races in these nimble craft. With the occupant sealed neatly in the seat, the kayaks were water-tight, buoyant, and extremely seaworthy. Usually single-place, they sometimes had tandem cockpits, for two. Out of the water they were always stored well above the ground on wooden supports, to keep them away from the ever-hungry dogs, who would happily make a meal out of their skin covering.

A young Inuit resident of "Pov" (Povungnituk) poses for Jeff Wyborn. The tent in the distance was a summer replacement for the igloo. Kayaks were stored well above the ground to prevent hungry sled dogs from eating their skin covering. *J. Wyborn*

It is a long time since I have been in the Arctic, and I don't know if these graceful craft are still used. If they have become a thing of the past, it is a shame.

The same may be said of another Eskimo masterpiece, the igloo. Everyone knows of these dome-shaped houses, which were built very quickly from blocks of hard snow. A little-known feature of some igloos was a slab of snow raised vertically on the roof to catch the fading horizontal rays of the sun and reflect them through a sheet of ice into the igloo itself. In this way the last natural light, and warmth, could be saved. In settlements such as Port Harrison, the igloo (its summer counterpart, made of canvas over a similarly shaped willow frame, was used at Rupert House) has given way to conventional houses. Maybe they are still used for overnight protection on hunting and trapping expeditions. I hope so.

As we flew over this extraordinary country there were many things to remind us that it was one of the most storied regions of Canada. It all began with the search for the North-West Passage and the fur

An Inuit igloo at Port Harrison. The vertical slab of snow is precisely located to reflect the sun's last horizontal rays down through a sheet of ice into the interior. Dips in the entrance tunnel trapped cold exterior air and prevented it from entering the igloo. *J. Wyborn*

trade—stemming from the popularity of the beaver hat, centuries ago, and then, more recently, the fur coat. British and French trading factions, often with military support, fought over the area, and whalers ventured into it in search of their prey. The settlements we were servicing often had far more history than the vastly bigger cities farther south. For instance, Rupert House had been established by the Hudson's Bay Company around 1668, Moose Factory in 1671, Fort Albany in 1679, and Captain James had visited the Charlton Islands in 1632, and had James Bay named after him. In 1904 a Hudson's Bay Company depot had been established in the Charltons, where ships from England carrying supplies for the James Bay region could be unloaded. In 1931, Charles and Anne Lindbergh had landed at Moose Factory while on their flight *North to the Orient* (the title of their subsequent book).

Jeff Wyborn at the controls of Austin Airways' Norseman, CF-GMM, which he flew for several thousand hours, proudly displays his Native-crafted mitts. *J. Wyborn*

THE PHYSICAL COST OF NORTHERN FLYING

I would like to conclude on a note of (faintly) humorous reality by mentioning the three occupational hazards that eventually plagued many bush pilots and mechanics of my era. While they may not all admit to these complaints, very few have been able to avoid them completely. They were the result of a career devoted to muscling heavy freight around the cramped cabin of an aircraft—stoop labour at its worst—and untold hours of sitting in draughty cockpits, often in damp clothing. In spite of Bob Noorduyn's best intentions, a lot of cool air always sneaked into the Norseman, around the pilot's legs. The ailments I am referring to? Lower back problems, arthritis, and hemorrhoids (piles!). Like everything else I've mentioned, they were a part of bush flying.

Whether pilot or engineer, we were all human, and had our aches and pains. Occasionally we became tired, fed up, and frustrated—but we always managed to bounce back, to see the funny side with a joke and a good laugh. That was our secret.

Lightnings over the Yukon

Flying a World War II Fighter on High-altitude Photography

ROBERT H. (BOB) FOWLER, OC

Bob Fowler retired in 1987 as de Havilland Canada's (DHC) chief engineering test pilot. His flying career was a distinguished one, earning for him both the McKee Trophy and the Order of Canada. He learned to fly at Toronto's Barker Field in 1940, joined the RCAF in 1942, and was eventually posted to 226 Squadron of the RAF, where he flew 48 tactical bombing missions on North American B-25 Mitchell medium bombers. After the war he returned to university, intent on a law degree. However, in 1947 the urge to fly got the better of him, and he became a pilot with Dominion Gulf, flying a Grumman Goose and Consolidated Canso amphibians on aerial magnetometer mineral survey operations. Two years later he joined Spartan Air Services, initially flying Canadian-built Avro Anson Vs on similar work. This is where he begins his narrative.

Subsequently (in 1952) he would leave Spartan to accept a position as a test pilot with DHC. In the next 35 years at DHC he carried out first flights on such world famous DHC types as the Turbo-Beaver, Twin Otter, Buffalo, Dash 7, and Dash 8. He also did the initial test flights of Canadian Pratt & Whitney's immensely successful PT6 turbo-prop engine, which is used in the Turbo-Beaver and Twin Otter. He has spoken to the Canadian Aviation Historical Society on numerous occasions and is known for his knack of presenting technical information in an understandable and entertaining manner. Bob

Fowler has the ability not only to bring his reader into the cockpit, but also to hand him or her the controls.

The subject of this story—the Lockheed P-38 Lightning—was one of the more famous Allied aircraft of World War II. It was designed as a twin-engine, long-range fighter able to accompany U.S. Army Air Force bombers deep into Germany. The Lightning's range, high-altitude capability, and high top speed also made it ideal for photo-reconnaissance (PR) missions to the most remote corners of occupied Europe carrying a battery of cameras instead of guns. It was this PR variant that served with Spartan Air Service as a successful high-altitude photo-survey aircraft—a most unusual "bush plane."

Not all pilots were capable of the extremely precise flying needed for aerial photo-survey work. The camera aircraft was flown at a "tape line" altitude of 35,000 feet on a "photo line," at the end of which it made a 180-degree turn and returned along another such line exactly parallel to the first one. The separating distance between the two lines was predetermined to allow for an overlap of the photographs being taken. An "intervalometer" controlling the camera shutter and keyed to the aircraft's ground speed (as opposed to air speed) insured that the same picture overlap occurred along the line of flight.

The camera was gimbal-mounted to remain perfectly level. The prints developed from the exposed film were assembled into a photo-mosaic from which cartographers produced the highly detailed maps that we take for granted. The efficiency—and thus the profitability—of the operation was directly related to the number of separate exposures required to cover a given area and the ground speed. Fewer photos meant lower costs. One picture from 35,000 feet covered 100 square miles. And the Lightning cruised at just under 400 mph true air speed.

Spartan eventually replaced its P-38s with de Havilland Mosquitoes that flew slightly lower but had more range and were easier to maintain. A competing firm used converted four-engine Boeing B-17 Flying Fortress bombers at a still lower altitude, compensated for by the Fort's even greater endurance. Bob's story appeared in the Fall 1991 CAHS *Journal*.

Dawson City, Yukon, June 1951—one of Spartan Air Services' Lockheed
F-5 (P-38) Lightnings, adapted for high-altitude aerial photography, is
prepared for a flight. *W. P. Doherty*

My P-38 flying with Spartan Air Services extended over one short year. No sooner had it begun then it was over; but it stands out in my memory. It was the beginning of my aviation education in a technical sense. Up to then I was pretty much a "knob jockey"—I enjoyed climbing into aeroplanes and flying them. Beyond knowing what to do when something went wrong (and I would hate to pretend that I always did), my mechanical interests (systems interest, if you like) to that time were limited. It was difficult, though, not to become fascinated with the hardware of a P-38 Lightning. Many of the systems were electrically or electro-hydraulically operated. But those lovely automatic gadgets went on the blink fairly regularly. When they did, it was good to have some idea of what was on the other end of the cockpit control.

Before discussing the aircraft, I would like to mention "Weldy" Phipps. He was the moving force behind the introduction of the P-38s into Spartan's operation. When the company decided to acquire a Lightning, "Weldy" designed most of the modifications and completed the bulk of the structural work and welding. He came up with ideas allowing the P-38 to do the job that Spartan needed, and I was somewhere well behind him. "Weldy" taught me how to weld, among other things. I enjoyed those days tremendously.

It was one thing to buy a used high-performance military aeroplane like the P-38, a very complicated machine, and quite another to make it do a high-altitude job in a civilian organization. I don't know of any piston-engine type from that era that was more complicated from a systems standpoint. "Weldy" flew the first Lightning to Ottawa without any training, or even the benefit of a flight manual. The aircraft wasn't in condition to do much of anything, much less the work that Spartan had in mind.

The P-38s that Spartan operated were U.S. Army Air Force (USAAF) F-5s—not the more common fighter version. Strictly photo-reconnaissance aircraft, they were capable of carrying up to five cameras in the nose, all operated by the pilot. These machines were flown during World War II at altitudes up to 40,000 feet, which seems incredible today. They performed a wide variety of photographic missions, one of which, called "tri-camera," was a type of multi-camera survey work, carried out over various parts of Europe for target-mapping purposes. Spartan obtained them in this configuration and modified them to place a navigator in the nose with one vertically mounted Fairchild aerial camera. A viewfinder continuously clocked ground speed in order to update the intervalometer. Knowing tape line altitude, it was possible, with a stopwatch, to establish ground speed fairly accurately. The navigator could then set the intervalometer to produce the desired line overlap between exposures.

Phipps also designed and built a new type of interior for the forward nose section, incorporating an optically flat window over which a specially designed Spartan drift sight could be mounted. There was also room in the nose for spare film magazines. A deck, on which the navigator lay prone while on survey, could be folded into a seat (with harness) for takeoff and landing.

Spartan bought its first P-38 in the U.S. in early 1950 and ferried it to Ottawa for modification that winter. It was registered CF-GSP, and, when the work was done, "Weldy" wheeled it out of the hangar and checked himself out in it at the original small Uplands airport. He had never flown anything quite like a Lightning. His only heavy aeroplane previous to the Lightning had been the lumbering Canso. And bear in mind that this was at the old Ottawa Airport (Uplands, prior to enlarging), where the longest available runway was a mere 3,600 feet. For the Lightning, that was a little on the short side. That he had no problems says something about "Weldy's" natural talent as a pilot.

That aeroplane did some early survey work in 1950, during which it dropped into Peace River, where I was based doing magnetometer work with an Anson 5. "Weldy" came in for fuel. He couldn't get 100/130-octane gasoline, and had to go on to Grande Prairie. But we had a good look at his aeroplane. I was intrigued. When I first sat in that P-38 I felt like the guy who could win the war single-handed—if someone would only make him a fighter pilot!

By the end of that year, the survey potential of the P-38 was more fully appreciated. Given good weather conditions, it was clearly capable of photographing tremendous areas in rather short periods. During this initial time, one engine thrashed itself to smithereens when the turbo-regulator failed. The many other smaller problems encountered gave Spartan some idea of the range of situations they could expect when the aeroplane was introduced into the photo survey business.

Two of the converted D.H. 98 Mosquitoes with which Spartan Air Services replaced its Lightnings, seen at Spartan's Uplands base (Ottawa) in 1962 after their retirement. *E. P. Gardiner*

The Canadian Government, at that time, came out with a program to complete coverage of the unmapped areas of mainland Canada. Much of the country was well covered from a map standpoint, but there were still great blank areas, particularly in northwest British Columbia, the Northwest Territories, and northern Quebec. These were the first places to be tackled.

Obviously, the greater the altitude the photos are taken from, the larger the area that could be covered per shot; and the faster the aeroplane, the better the production rate. This naturally led Spartan to consider something like the P-38, which was capable of attaining a high altitude with a respectable level-flight air speed, at 35,000 feet. One exposure with the cameras we were using covered an area of 100 square miles! This appealed to the government because it reduced the number of prints required to produce a controlled map.

Our second Lightning, registered CF-GSQ, arrived from the U.S. later in 1950. Having seen the first one, I mentally put my brand on this one. After it arrived in Ottawa, I tried to stay within about three feet of it—even while eating my lunch. "Weldy," Bill Doherty, Len McHale, and I began work on the two of them in preparation for the 1951 season. Both aeroplanes were extensively modified during the winter of 1950/51, and then "Weldy" and I took them to Dawson City, Yukon, in the first week of June 1951. "Weldy's" navigator/camera operator was Benny Benoit, a veteran of RAF Bomber Command, while I had Joe Kohut DFC, who had been in the RAF's Pathfinder Force.

The only windows in the nose of these machines were little triangular panels on each side (ports for oblique cameras), and a large square one on the bottom. To remedy this, "Weldy" glazed part of the nose-cone. He took the metal nose-cap off, cut the centre out of it, and made a frame inside. Then he moulded some Plexiglas, and trimmed it to fit. This gave the navigator a forward view.

There were radiators on each side of the P-38's distinctive twin tail-booms. You can imagine the lengths of glycol coolant line that flowed back and forth in the aeroplane between these and the Allison engines. The aircraft was full of pipes, ducts, and wires; everything that made it run made long journeys fore and aft. A small intake on the top of each boom provided cooling air for the turbine-feed that drove each of the turbo-chargers, which were flush with the top of the boom. The exhaust gases from the engines drove the turbines.

Windows that slid up and down enclosed the cockpit on the sides. Little cranks, much like those used in a car—except that they were fitted with ratchets—opened and closed the windows. A transparent overhead canopy section, hinged at the rear, was latched to the top of the windshield at its forward edge. The side windows were somewhat complicated, but they sealed up quite well if you took the trouble to get them tight. The armoured windshield was a good two inches thick, a typical fighter feature, and the cockpit was roomy. The wheel seemed to come to hand very comfortably, except for a slight forward reach. To operate the wing flaps the pilot had to change hands on the wheel, which was a nuisance.

Everyone who flew the Lightning enjoyed the experience. It was smooth, quiet, and very pleasant, when everything was working right. We never missed an opportunity to do low passes—always discreetly—to get a good look at a windsock or for other highly *essential* reasons.

Joe, my navigator/camera operator, had sustained some flak damage during the war, which left him with one lung missing—and here he was pressure breathing at 35,000 feet for several hours almost every day. It never seemed to bother him. (With the pressure breathing of oxygen, after the pilot had exhaled, his first inhalation triggered the delivery of oxygen under pressure from a tightly fitted facemask. Exhalation stopped the process.) Benny Benoit, in the number one aircraft flown by "Weldy," was a very aggressive navigator. Our two supporting engineers were Bill Doherty and Len McHale.

One pump per engine supplied the hydraulic system. If you sprung a leak, you lost everything; with no fluid to cool them, both pumps would fail. Fortunately this did not happen frequently, but most P-38 pilots experienced it at least once. All hydraulic system services would function with one failed pump. Early aeroplanes had a single DC generator, but we had one on each engine. The less said about the electrics of the Lightning the better.

An intensifier tube system, much the same as in a Beaver or an Anson 5, heated the cockpit. The tube, surrounded by exhaust gas, heated outside air before it was ducted into the cockpit. Although there was lots of warm air, it seemed to be directed at specific places—everything else froze. All manner of draughts in the cockpit entered mainly through small cracks in the side windows and canopy. With the outside air temperature at or close to -60 (F), a hand or some other part of the anatomy could suddenly begin to go numb. Before you realized that the window on one side was not completely closed, your hand could become almost useless with the cold. At the same time the heat could be melting the rubber of your flight boots—without necessarily keeping your feet warm.

The toe brakes (used for ground taxiing) were not power-operated. They functioned much the same as those in a car. After these brakes had been set up, they worked beautifully for about three flights, and from then on it became a steadily increasing pumping operation. If you really want to build up the muscles in your lower legs, I recommend taxiing a P-38.

Also worthy of mention was the poor visibility from the cockpit. On a clear day a pilot could certainly see lots of P-38 in any direction—left, right, or forward. And to the side, and even ahead to some extent, there was a wonderful view of the top of the wing. Through the windshield there was an equally impressive view of about eight

Breathing oxygen, pilot Bob Fowler sits in the cockpit of Lightning CF-GSQ at Dawson City in 1951, in readiness for another survey flight at 35,000 feet. *Via R. H. Fowler*

feet of nose. If you peered farther out along the span, you saw the booms stretching aft on both sides, obliterating great chunks of countryside. It must have been tough for pilots who flew the Lightning as a fighter to know what was going on around them. It would have been necessary to keep the thing constantly rolling in order to open enough sky to see the enemy. They could have come in from any direction, and the pilot would have been the last to know about it. The best visibility was forward over the wing leading edge, which was ahead of the cockpit by a good three feet. Even that view was not the greatest.

The rudders had to be operated manually. Each rudder was bathed in propeller slipstream, and on takeoff, due to the opposite rotation of the propellers, you could almost keep your feet planted on the floor. The aeroplane went as straight as a die down the white line as though it knew what it was doing—until an engine did something impolite. Then those relatively pleasant rudder forces became very heavy. If you lost an engine on takeoff in a P-38, the leg on the rudder pedal that maintained a force against the good engine (to counteract the absence of power on the other side) would remind you of the event for some time to come. Acceleration on takeoff was impressive. When you released the brakes, with a total of 3,400 hp, the aeroplane really moved down the runway.

The ailerons were also hydraulically assisted. With the boost on, it was easy to do lovely rolls with one hand. You simply had to cock the nose up a bit, put the wheel over, add a little inside rudder, and it would do a roll with the pilot firmly in the seat. The aileron boost made handling very pleasant. If for any reason you lost hydraulic pressure, you had an aeroplane that required two hands and hockey gloves to do a turn: it felt as heavy as a Canso or a Lancaster.

Both machines left Ottawa in early June, each with the two untested 165-gallon long-range drop tanks hung underneath. We barely had time to put the tanks on, fill them, and take off for the west. We had never tried them at altitude; in fact we had never tried them at all. But there was a great urge to get moving, and, if there were a problem, we would just take them off. On the way out, at 20,000 to 25,000 feet, they fed well.

Most of our cross-country flying was below 25,000 feet. In those days, you had the upper altitudes pretty much to yourself except for Don Rogers (former Avro Canada test pilot), who used a fair bit of them with the (Avro) Jetliner (the first civil jet transport to fly in North America). Our P-38s were very nice for going over weather with navigation mainly by ADF (automatic direction finder, which homed in on radio beacons.) From Ottawa, we headed west to Winnipeg. In the morning we flew on to Edmonton, and spent the night there. Then we headed up to Dawson City with a stop at Whitehorse on the way.

There were two strips to choose from at Dawson, which confused things at times. More than one aeroplane came a bit of a cropper at the wrong strip. A pilot named Pat Callison had a strip some 1,600 feet long and no more than 20 feet wide at Bear Creek. The aspect (length to width) ratio of that strip made it appear to be 6,000 feet long when it was viewed from 2,000 feet. We circled it, and something about it didn't look quite right. Happily we found the Dawson strip, which was turf, a good 100 feet wide and about 3,600 feet long. Yet it did look shorter than the one at Bear Creek.

While we were in Dawson, some American *Flying Doctors* who were on their way up the Northwest Staging Route to Alaska managed to land on the Bear Creek strip. At least one went off the end and broke an undercarriage leg. It was the old aspect ratio effect that caused him to land at the Callison strip. Typically, "Weldy" did a lot of the work, and got the doctor's little Bellanca Cruisair fixed up to the extent that he could fly it back to Colorado with the gear down.

We soon found that we could not operate satisfactorily at the 35,000-foot altitude necessary for our photography when we were carrying our long-range fuel tanks. This was sobering. Without those extra 330 gallons, our survey capability out of Dawson was rather limited. We were left with an endurance of two-and-a-half hours and a bit. Subtracting climb and descent time, plus 20 to 30 minutes reserve, gave us an outside maximum of one hour and 40 minutes "on line." On subsequent flights, after takeoff from Dawson, we would climb to the north, and as soon as we hit 35,000 feet we surveyed to the Arctic coast and back again to a carefully planned descent point for the letdown into Dawson. Fuel was very carefully monitored. At 35,000 feet, indicated air speed was 210 to 220 mph, which gave us a true air speed (TAS) of 370-odd miles per hour.

The part of all this that most interested us was the 27 cents per line-mile bonus we were getting. The little cash registers in our heads worked at a great rate, totalling up 370 times 27 cents for the time we spent on each photo line. We all had plans for spending that mileage bonus at the end of the season.

At altitude we frequently encountered a fair bit of clear air turbulence (CAT), which invariably made it difficult to hold the aeroplane level just before the exposure light was about to go on. At times, CAT above 35,000 feet could be quite violent, making acceptable level flight almost impossible, and many times we were concerned that the pictures we were taking might turn out to be useless.

About the only other routine weather phenomenon that often caused us to pack it in was cirrus cloud in extensive sheets between us and the ground. It always seemed a little odd to quit because of a fairly thin layer of cirrus; but of course it could completely spoil the photography. Wind, as such, was not much of a problem at 35,000 feet—except when breaking it! Then it could become a little more complicated than one might anticipate. The less said the better.

To maintain a consistent scale between pictures taken on different flights, it was important that each flight be flown at exactly the same height above the ground. We found that haze was a problem in the valley at Dawson, where the strip was located. Good weather for an extended period had resulted in numerous forest and brush fires. Below 7,000 to 8,000 feet, the haze was quite thick. The strip at Dawson was in a fairly deep valley, with 1,800-foot hills parallel to the strip on two sides, which effectively prevented a proper circuit. Of course this was a fine excuse for a low pass along the strip—for a look at the windsock, which couldn't be properly seen from above the hills.

Typically, we located the valley by coming down the Yukon River until we were within sight of Dawson City. Then we dropped the landing gear, turned into the valley, and groped our way down, straight onto the Dawson strip. If the wind was wrong, we usually flew down the valley, turned 180 degrees at a wider space, and did an approach from the other direction.

With 24 hours of daylight, we had plenty of time for survey flying; but we had to call a halt in the late afternoon when the shadows on the ground grew long, hiding detail in the river valleys. Another drawback of 24 hours of daylight was the difficulty we experienced in getting a good *night's* sleep, particularly with the noise from fracases, which would develop outside at any hour of the so-called night. It didn't take much imagination to picture those old sourdoughs ram-

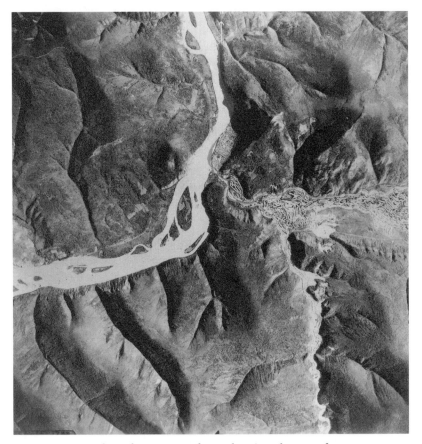

Dawson City, Yukon, from 35,000 feet, taken in July 1951, from one of Spartan's Lightnings. The strip from which the aircraft operated is in the midst of the pattern of dredge tailings to the east of the town. *Spartan Air Services*

bling around the town. We lived at the Wellington Hotel, one of the many buildings that were slowly sinking into the ground as they melted their way into the permafrost.

People in Dawson were extremely friendly. It seemed they couldn't be kind enough to visiting strangers. We really enjoyed ourselves. We visited the cabin where Robert W. Service did much of his writing while he worked at the Bank of Commerce. On one wall of his cabin were the words, "Don't worry, WORK" scrawled in crayon long ago.

At the Dawson strip we refuelled the Lightnings from drums brought up from Whitehorse on a huge barge that was pushed downstream by one of the sternwheelers still running between Whitehorse and Dawson. The fuel, delivered to Dawson, cost Spartan about $1.25 a gallon, which in those days seemed atrocious! In any case, it still worked out well considering the sort of money the Canadian Government paid to photo-survey that particular piece of the Yukon.

The weather was good for about the first six-and-a-half weeks; then it began to change for the worse. When we couldn't operate regularly, the toss of a coin decided that Joe and I would fly down to Vancouver to tackle another area Spartan had bid on in southern British Columbia. Len McHale followed us down on CPA, and we set up shop at Vancouver Airport.

The area in British Columbia to be photographed was south of an east-west line roughly between Comox and the Alberta border. Overall, it was similar in size to the Yukon area. Again we had excellent weather and flew six hours a day, almost nine hours on some days. Between the two first flights of the day (about three hours each), typically we landed, refuelled, and picked up full film magazines. Joe Kohut and I usually stayed in the aeroplane breathing 100 percent oxygen while we refuelled. Later we realized that this was very foolish. We were beginning to suffer from the bends with increasing regularity now that we had been at the job for two months.

Early on one memorable flight on a photo line, after we had been in Vancouver a few weeks, I suddenly felt as though a spike had been driven into my left knee. This began to happen regularly. When these "big" bends occurred, we had to land. There was no point in trying to stay at 35,000 feet. We had tried coming down until the greatest amount of pain subsided, usually at about 25,000 feet, and then climbing back up to 35,000 feet to resume the line. We wouldn't be there more than about five minutes before it recurred.

Looking back, there were many aspects of sitting at 35,000 feet almost every day, puffing away at the old pressure breathing—actually intended as an emergency measure in pressurized fighters—that were unique. Except for the odd occasional course change, Joe and I sat there for extended periods with almost no communication between us. When we remembered, we would check in with each other to be sure that nothing had gone wrong with our oxygen systems. Cold was a constant problem, and likely contributed to the increasing incidence of bends.

While "Weldy" had begun to suffer with the bends quite regularly at Dawson, I had experienced only the occasional twinge. I began to think that I was pretty tough because I was not feeling the same degree of pain. It hurt him so much, it made his eyes run. He could not sit still and would have to land. The bends would eventually end my high-altitude flying in the Lightning.

Another aspect of this rather lonely job was the sheer monotony of sitting in that big transparent cockpit with the tremendous glare, just listening to one's own breathing. Flying in one direction, wherever the sun struck you, you were nice and warm, but everywhere else you were as cold as ice. When we did a turn-around, and reversed direction, our warm and cold parts were switched.

I would only occasionally gasp a few words to my partner in the nose. Pressure breathing is a tiresome business, and doesn't do much for intelligible conversation. I found an unused instrument hole near the top of the panel through which I could see the rear bulkhead of the nose compartment. I cut a two-to-three-inch diameter hole in the bulkhead, in line with the empty instrument hole. From then on it was a whole new ball game. Through that little hole I could see Joe down in the nose, moving about and doing his job, and while it was a bit like looking down a pipe, it made a world of difference, and time passed much more quickly. It was comforting to know that we could then exchange notes if something went wrong with the intercom.

One final comment on the physical side of the operation—it was a great way to lose weight! All of us were as skinny as rails. I noticed this in a picture of Joe Kohut and me taken on the street in Vancouver. We were almost unrecognizable. Too bad I can't produce the same effect today.

I finally finished the season at Vancouver after putting in about 158 hours flying in just under seven weeks, almost all of it at 35,000 feet. Spartan's first full season with the two aircraft had gone extremely well, and I believe we produced the most revenue that the firm had seen in any one year up to that time.

But the bends persisted, and I began to have dizzy spells after I got above 32,000 feet. The old head would go around, and it was difficult to stay on top of things. I usually had to drop down to about 26,000 feet before the effect would disappear, leaving me with a fine headache. After a few flights like this, I began to wonder if all was well.

One of the Lockheed F-5B (P-38) Lightnings modified by Spartan
Air Services, probably CF-GSQ flown by Bob Fowler, photographed at
Vancouver in August 1951. *D. E. Anderson*

So I went to see a Dr. Cooper who was the medical officer for CPA at
the time. When he heard of the number of hours we were putting in
each day breathing 100 percent oxygen, both in the air and on the
ground, he said that there was no mystery here. We were upsetting the
balance of gases in our bloodstream, and this tended to upset our
vestibular system. The best thing, he said, was to stop for a while, and
stop I did. That was when I left high-altitude operation at Spartan.

Canso Aerial Mineral Surveys

Contour Flying at 500 Feet with 16 Tons of Aircraft

JOHN ("SMITTY") SMITH

The following account is based upon a presentation given at the 1993 Canadian Aviation Historical Society Convention held in Regina, Saskatchewan. At the time, John ("Smitty") Smith, a long-time member of the CAHS, was vice-president of the Toronto chapter. In January of 1994, he underwent minor surgery to repair a shoulder injury (a torn rotator cuff) sustained a few months earlier in an industrial-league hockey game. During the operation "Smitty" lapsed into a coma from which he has never recovered. A transcript of his talk was adapted for the Winter 1995 *Journal*.

"Smitty" was 17 when he first soloed in 1948 on a Fleet Canuck belonging to the Toronto Flying Club. An air cadet flying scholarship had given him 12 hours of dual instruction, and he would put in a further five hours solo. To build flying time he had negotiated an arrangement with Bruce Best, the club's flight engineer: four-and-a-half hours of sweeping the hangar floor translated into a half-hour in the air. "That was the beginning of my long climb, flying first as a private pilot and eventually commercially," said "Smitty." As well as the Canuck, he flew the club's two Tiger Moths and "built up quite a bit of time, thanks to Bruce."

"Eventually, by begging, borrowing, and stealing, I became a '200-hour wonder.' With my flying time behind me and my commercial ticket, I joined Kenting Aviation on Avro Anson Vs." One of the firm's veteran pilots, Charlie Parkin, encouraged him and gave him his twin-engine rating. "I was young and really didn't know a helluva lot. Charlie was a gentleman with a lot of patience, and I owe him a great deal." His qualification on twin-engine aircraft permitted "Smitty" to fly Kenting's Cansos, which are the subject of his story. He would later check out on the firm's four-engine Boeing B-17 Flying Fortresses used for aerial photography—a long way indeed from the tiny Canuck on which he had soloed.

As "Smitty" explains, the Consolidated PBY-5A Canso was a formidable aircraft, and certainly an unusual bush plane. With a span of more than 102 feet and length of almost 63 feet, it was comparable in size to the DC-3, but several tons heavier. During World War II, it flew with no fewer than seven air forces, including the RCAF, mostly in an anti-submarine role. It operated with considerable success, and its crews were credited with sinking many subs. Flight Lieutenant David Hornell won the Victoria Cross while pressing home an attack on a U-boat in his Canso. An example in the National Aviation Museum bears the markings of Hornell's aircraft. Although it was an American design, Canadian Vickers and Boeing of Canada built it.

In his story, "Smitty" tells us about airborne geophysics and the Consolidated PBY-5A Canso amphibian—about survey flying at 500 feet with a 16-ton aircraft. His employer, Kenting Aviation, operated across Canada and abroad, carrying out mainly mineral surveys.

Let me take you with me into the cockpit of a Canso for a typical day of flying with Kenting Aviation. While my scenario could be anywhere in Canada, in this instance we will assume Kapuskasing, Ontario, to be our operating base. But it could as well be Val d'Or, Lethbridge, Moncton, or Chibougamau. Our typical flying day begins with a 6:00 AM wake-up call from the hotel desk. I roll out of bed and immediately phone the met (meteorological) office to check out the weather in the survey area and find out what is moving in from the west. I am trying to figure out what we can expect during the day. I learn that the weather will be suitable for flying, so I call each of my crewmembers. Waking them up, I cheerfully inform them when breakfast will be served and when the cabs will leave for the airport. After our meal we pick up sandwiches from Mary in the dining room, and assemble in the lobby. The only food we take is what we can carry in our jacket pockets—a sandwich and a Coke. These will keep us going for seven hours under some pretty tough conditions. Bumping along at 500 feet, it's hard to eat a sandwich, let alone drink a Coke.

At the field, I meet our flight engineer, who briefs me on our aircraft, the amount of fuel on board, and what to look for as far as maintenance checks. He also tells me what he plans to do to the aircraft when we return to prepare it for the next day.

The magnetometer technicians check out their equipment while I pull myself up on the nose by the bow rails on the front of the Canso. Stepping over the windscreen, I take up a position just aft of the cockpit to assist the flight engineer. We pull the props through to help disperse any oil that might have built up overnight. This is a daily routine. Then I climb on top of the wing, using a step set into the front of the tower that supports the wing, and walk out to the tips to see if the retractable wingtip floats are firmly locked in the "up" position.

I wear shoes with soft rubber soles. Climbing on the smooth contours of the nose requires sure footing. I also go out on the wing—some 10 feet above the ground—and jump up and down to

Kenting's Consolidated Canso CF-IKO had its nose turret replaced with a Plexiglas dome. *R. Pettus*

170

check the floats. An old pair of rubber pants and a flight jacket complete my outfit. Under that glazed canopy all day long, I will perspire freely, and the fleece lining sops it up. And of course I have my Ray-Ban sun-glasses. I am 22 years old and flying a Canso. Can you blame me if I want to look cool?

Our pre-flight checks completed, we fire up our two 1,200 hp Pratt & Whitney Twin Wasp engines and taxi out for takeoff. Once in the air, we circle back and do a low pass, along the runway, to check out the accuracy of our radio altimeter. We fly along the runway at 500 feet and set the altimeter to indicate zero feet. That setting will guarantee us 500 feet of altitude—the height at which we will be contour-flying all day. The altimeter registers in 100-foot increments—and they pass so quickly! We may be flying shallow contours, very steep contours, or even over mountainous country, but it makes no difference. It is vital to maintain precise height because we will be sending a signal of a certain strength into the ground and measuring it on its return by means of a "bird" trailing behind the aircraft. Our altitude will be recorded continuously, so there can be no cheating. The technician will tell us if we are too high or too low; he can see it registering on his charts. Regardless of terrain, maintaining that 500 feet is an absolute necessity.

By the time we approach the survey area, the technician has already lowered the bird from its cradle under the aft part of the fuselage—and it is flying beautifully. Sometimes a bird won't fly properly, causing a lot of drag. At other times, as on this flight, it tracks faultlessly behind the aeroplane. We are beaming a constant electrical signal into the ground, and the bird collects anything that bounces back. A fibreglass pod about five feet long and 10 inches in diameter with stabilizing fins, the bird is full of receiving coils that measure any readings—deviations in the Earth's magnetic field—that come back from the ground, should there be a conductor there in the form of an ore body.

If there is no buried conductor below us, there will be no readings on our charts. We run two sets of these, at a location just aft of the pilot's compartment. The magnetometer is positioned at the rear of the fuselage, aft of the tail, to keep it from being interfered with by the engine magnetos. The "maggie" needles and other electronics are very sensitive.

John ("Smitty") Smith as a very youthful co-pilot beside one of Kenting's Canso aircraft. He has a set of pilot's wings sewn on his high school football jacket. John always kept a football with him in the aircraft on the chance that he might find someone to catch a few passes when he landed. *Via Elinor Smith*

At operating altitude (500 feet) holding a course on my directional gyro (DG), the navigator gives the order to "start cameras." We also have cameras running constantly—two in case one fails. He natters continuously, giving me corrections, "right" or "left," as he takes us down our line of flight. Picture a map with straight grid lines all the way across. The navigator doesn't take us from point to point; he is dead reckoning on a straight line. It is an overcast day, and, being early in the morning, it is easy to hold a heading. With a heading on my DG and, luckily, stable wind conditions, I line up a row of rivets on the nose with an object way down the line and just hold to that heading— for maybe three, four, five minutes. It is gratifying to know that I am tracking the aeroplane exactly where the navigator wants me to go, by picking out known points on the line, with no need for corrections. Since it is so important to hold that heading, I am pumping those rudder pedals a lot—you can develop muscular legs holding a 16-ton aeroplane on such a precise line. It is hard work, and youthful stamina is an asset. It also helps if you love flying and know your aeroplane well. But it is so demanding that some pilots have trouble with this type of flying.

We have to be very patient with the navigator. The poor guy is in there for over seven hours without a break, since we just fly and fly and fly. At such low altitude this is no easy task for any navigator. Rarely do we have to break off a line and climb so that he can reorient himself. But should this happen, we immediately drop down again and continue our flight pattern. While we two pilots do an hour on and an hour off—one flying the aeroplane while the other handles the power—the navigator's job is nonstop.

Our navigators are very talented men. On the west coast of Hudson Bay, around Rankin Inlet—to give you an appreciation of how difficult it is for a navigator to "dead reckon"—it is very flat, like a vast white sheet with absolutely no features, even as low as 500 feet. Yet our tracks register perfectly straight. The navigator, who sits behind us working away with his "top" (topographical) maps rolled up on his lap, has done an excellent job.

Thus we spend a more or less typical day, flying for the seven or so hours that our fuel supply permits.

CHALLENGES WERE ALMOST ROUTINE

Some jobs we did in shifts, getting up very early at first light and flying. Then we would come back, and the second crew would take our place. I've actually done two shifts of five-and-a-half hours—11 hours of flying in one day—just to get a job done.

Leaving Kapuskasing we once flew out to Cranbrook, British Columbia, for an assignment over some very mountainous territory. Flying across valleys at 500 feet, we would be staring straight into the middle of a mountain, wondering how in hell we were ever going to get up and over it. As we approached the base of the mountain, the pilot would raise one finger for an increase in power, then a second finger for more power, and finally a third for maximum climbing power! As we struggled to claw our way up the face of the mountain, we soon realized there was no possible way we would ever get to the top. Even though we were flying at maximum power, there was still a lot of mountain above us. So we did what was almost a stall turn, which is not really as dicey as it sounds. We were experienced and knew exactly what we were doing. We knew our aeroplane and took it to its limits. We simply turned back down the mountain, climbed at a less acute angle, and approached it horizontally to pick up the line where we had abandoned it. Then we repeated the whole process. Or we flew the line in reverse, in a shallow dive from top to bottom.

Going to full power, we opened the cooling gills on the engine cowlings, allowing more air to flow through, and thereby lowering the cylinder-head temperature. When the throttles were chopped—to no power as we broke off the line—the gills were closed. Then back to almost full power—with the gills open—to keep up speed for the climb to the top of the mountain. Then back down again—with no power—as we plunged down the opposite side. Our engines were worked-over viciously. And so were we. Days like that—after seven or eight hours—just whipped you. But we managed to complete the survey.

At 500 feet, flying over all types of terrain ranging from rock, to water, to trees, to sand, to bare earth, we experienced extreme turbulence. All of these surfaces reflect different amounts of heat—particularly from noon on when the Earth's temperature is at its highest. The rapidly rising air created bumps that pounded us constantly. At the same time, we might have to contend with a crosswind.

Canso CF-DFB trails a bird over a spectacular but unfortunately unidentified landscape. *R. Pettus*

It was not unusual to crab down line, with the nose 20 degrees off to one side, just to hold the machine on track. Keeping 16 tons of aeroplane on a precise course under such conditions was never a picnic.

Could there be a more demanding type of flying? Certainly not, I thought. But another survey pilot, Bob Pettus (my captain when I first started fresh out of flying school), who had flown thousands of hours on both geophysics and water-bombing, assured me that water-bombing was just as strenuous.

For each trip in the Canso we carried 800 gallons of fuel, consuming it at 100 gallons per hour. While many pilots could not stay up for the full seven hours plus, we were often closer to the eight hours our fuel allowed us. We just flew and kept flying, determined to get the job done. Should we fly six days straight—pretty well the norm—we automatically took the seventh day off, regardless of aircraft serviceability, weather, or even pressure from the company. We were physically and mentally beat from pushing that aeroplane for all those hours. Relaxing back at base did the crew a world of good.

We never seemed to stay long in any particular area, perhaps a few weeks at most, before moving on to our next assignment. There were always other projects waiting. At one time, Kenting had four Cansos, all doing similar work. We were an active company. Our Cansos were registered as CF-GKI, 'IKO, 'DFB, and 'HVV.

THE CANSO

For its time, the Canso was extremely reliable and ideally suited for our work. It had great range and could carry a huge load—including its own spares and those of the technicians (and everyone carried back-ups). The weight and drag increased even more when we added the cable out on the wing and the bird under the hull. We operated only on wheels. Because of the weight of the equipment under the aircraft, we never flew from water on "maggie ops." In all the years Kenting flew Cansos from the North Pole to the South Pole and as far east as New Guinea—under the most rigorous conditions imaginable—we had no serious mishaps. From the 1950s to the 1970s our Cansos brought their crews home. We would look up at those engines and say, "Thank the Lord—and Pratt & Whitney!" The Canso—there will never be another aeroplane quite like it.

176

OCCASIONAL EXCITEMENT

Did we have any close calls? You bet. Sometimes we had water in our gas during northern operations—even though we pumped our own fuel through a chamois. This would cause a momentary loss of power. At 500 feet, that could be a thrill! One of the engines would suddenly go, "brrrrup, brrrup, brrup, brup!"—and on and off so quickly that we had no chance even to determine which engine it was. We didn't try to sort it out. We put the power up and kept straight, trying to gain altitude—buying time—and sitting tight with our big lap belts firmly buckled. Then we tried to pinpoint our problem. "Where the hell can we land?" was always in the back of our minds. This happened more than once. We came through, but each time we aged a bit.

I once had a piston valve that, for some mysterious reason, would malfunction in the middle of the afternoon—not on takeoff, not in the morning, nor at any other time. In the middle of the afternoon, the engine would lose power, as though someone had yanked off engine magneto cables. The engine went "Boom! Boom! Boom! Boom!" And then it would run smoothly. We lived with the problem for a couple of weeks, until our chief pilot, Jimmy Greenshields, arrived in a Lockheed 14 and asked for a flight.

So we took off. It was in the morning, early on our day's survey, and the engine ran smoothly. No problem; nice day. As it neared 2:00 or 2:30 in the afternoon, it was time for the engine to go. Jimmy and I were alternating at the controls, an hour on/hour off, and I hoped he would be flying when the engine acted up. Sure enough, he got it. No sooner had he grabbed the control column then, suddenly, "Boom! Boom! Boom! Boom!" He straightened up in the seat. "Oh!" he exclaimed. "I see what you mean!"

We returned to base, and he questioned the engineer. "Have you pulled the filter screens?" When a Canso's engine decides it's going to give up, you pull the screens and generally you find metal particles telling you, that's it—change the engine. We pulled the screens and found nothing. We had already changed magnetos and plugs and everything we could think of. So, my problem remained a mystery.

"Well, Jim, what's your decision?" I asked.

"You're the captain," he replied. "You decide."

"Is that aeroplane safe to fly?"

"I don't see any real problem, Smitty. I think you could finish the job."

"Are you telling me to fly the aeroplane?"

"Yeah."

But a few days later the engine acted up again, and continued to do so with increasing frequency. So, I returned to base and grounded the machine. "Check that screen again," I said. This time we found particles. So we changed the engine and resumed our surveying—with no further excitement.

Often we would practise single-engine emergency flying with the Canso. Flying at just five wingspans above the ground, our procedures needed to be pretty sharp. We had to handle any situation instantly. We could stop an engine and still keep the aircraft perfectly steady, almost unaffected by the loss of power. We had to be certain that everything was functioning—if we hoped to make it back to base.

We once set out from Bathurst, New Brunswick, on a copper survey for Texas Gulf. It was very early in the morning; we had a full load of gas and were just west of Bathurst down below the hills when the starboard engine blew a cylinder. We managed to climb above the hills on one engine and make for the nearest field, the Chatham RCAF Base in New Brunswick. We declared an emergency, and they cleared us.

Five minutes after a forced landing at Chatham, New Brunswick, RCAF station (because of an engine failure), a crew readies Canso CF-DFB for an engine change. *J. Smith via E. Smith*

But immediately upon landing—right there on the runway—we were surrounded by military police vehicles. The MPs came out with their guns drawn and ordered us out of the aircraft. We were thoroughly interrogated right then and there. They thought we were attempting to test them by breaching their security—even with our feathered prop! Only when they were satisfied that our plight was real did we receive their cooperation. The next morning one of our B-17s flew down from Oshawa, Ontario, with a spare engine, and we got to work on it. Such incidents were a way of life. When something broke, we simply replaced it and advised the insurance company.

I have distinct memories of one particular winter takeoff at Churchill, Manitoba. We were heading up to Rankin Inlet in the Northwest Territories to do a nickel survey. Whenever we raised or lowered the landing gear in the old Canso, we could sure hear it. It made one incredibly loud series of thumps. One wheel went "boom," then the other, and finally the nose wheel banged into place. This time there was a lot of unusual vibration as the nose wheel retracted into its housing between the two pilots' seats. The housing had a little window and, when the nose wheel came in, there it was. The vibration stopped abruptly.

As I've said, we became pretty finely attuned to the aeroplane after flying it under such conditions for so long. I felt that my fingers had almost become part of the wing-tips. So I decided that, while something might be wrong, it wasn't worth worrying about it. We pressed on up to Rankin Inlet, where we flew for the usual seven hours looking for nickel. When we came back we joined the circuit and lowered the wheels. "Boom ... boom"—the starboard wheel came down, followed by the port. But no third thump. The window in the nose wheel housing remained full of black rubber—the wheel had not lowered properly.

We called the ground engineer and were advised to use our prybar (part of our emergency equipment) to push the wheel down. We did so. Once we got it clear of the housing, it dropped and locked into place. It really wasn't much of a problem—luckily. We were very low on fuel and had begun to wonder how we were going to get down without damaging the aeroplane. Landing on the nose would have meant major hull damage. Later, we learned that the problem originated with a worn nose wheel, which had been replaced with a new one that was out-of-round. When it retracted, still rotating, it struck the sides of the well and jammed.

At times it could be damned cold—a really crisp -40 (F), when walking on snow was like walking on crackers. On such days you could swing on a prop and not move it. So, out would come the Herman Nelsons to heat both the engines and the inside of the aeroplane. But we still flew.

I recall some difficult winter contour flying south of Kapuskasing, Ontario, where the ground surface was made up of a succession of shallow depressions. I was flying at 600 or 700 feet—I couldn't safely go lower, as I explained to the chief technician.

"Well," he worried, "I don't think they'll accept the work we've done here."

"Okay, Derek, I'll try it again, just for you."

So, for half an hour, I tried to dip in and out, speeding up as we came out of each depression in a nose-high attitude, trying to maintain that 500-foot profile. As we slid down, the bird, trailing behind, was dropping even lower. And of course the occasional tree was a little taller than the rest. The bird was fitted with stabilizing fins, and on one of our dips it apparently snagged a tree, shearing the fins off and causing it to track erratically. After the impact, the technician came forward.

"I think we've got a problem," he announced.

"We break the bird?" I asked.

"Yeah, I think so."

"Go back and have a look."

"What have we got?" I asked when he returned.

"We got a bird with no tail feathers."

Of course, that ended our day's work. We had a procedure to follow. All our Cansos were equipped with double-bitted axes, and it was not unusual, if you had a bird problem, to chop it free, literally. We had to go back in the hull of the aircraft, look down through the hatch at the bird, and decide whether or not it had to be chopped free. Occasionally, pilots did lose their birds. Maybe they just got carried away, grew a little tired, or simply flew their contours too tightly.

In this case we tried everything. I made steep turns, trying centrifugal force to get that bird out of its slope. I did everything possible to shake it free and get some tension on it. But as soon as the bird

One of the fibreglass "birds" that were towed by Kenting's Cansos. *H. Humphrey*

came up under the hull of the aircraft, it would just lift, hitting the control surfaces and aggravating the problem. No matter what we tried, as soon as we got the bird anywhere close to the aeroplane, it would start oscillating violently, going wild.

"There's just too much chance of it wrapping around the tailplane," I told the crew. "We could lose the whole bloody thing, aircraft and all! We'll go back to camp."

We had about 10 feet of snow in the middle of the field, and I felt I could fly slowly across, wheels down, hanging on the props—with someone back at the hatch to guide me "lower—lower—lower" until the bird actually touched the snow. Then I would give the order, "Chop!" And this was what we did. But the bird didn't cooperate. It dove into the snow and came right back up again just like the bouncing bombs used on the famous Dam Busters raid: "Foom . . . foom . . . foom!"

"How far did it go?" I asked.

"About half a mile, Smitty. It was really flying!"

Next day I called headquarters. "Hey, send me a bird," I requested. They did so, and we installed it on the aeroplane and were right back in business.

Did we have any trouble with actual bird strikes? Oh, the odd eagle would come at the canopy, but the slipstream generally carried it up and over. Other times, they'd just square off in front of you.

Kenting did lose an Aero Commander and its crew of four at Thompson, Manitoba, while I was working out of there. We had two of these light twin-engine aircraft teamed up on a surveying assignment. The size of the job diminished to the point where only one aircraft was required, and so the second aircraft took off for home. It did a 180-degree turn and was coming back across the field—which was a bit unusual. It had just eased up, taking a heading for Winnipeg, when the starboard wing rotated upward. It didn't fall off. It just rotated when the spar failed, apparently from metal fatigue.

JOB SATISFACTION

I guess one of my two biggest thrills was flying the original Thompson, Manitoba, nickel survey back in 1953. That became a major Inco mine with a small town of its own, and even boasted an airport.

My second thrill, our biggest find, was in 1953 while I was based at Gore Bay on Manitoulin Island. From there we did the original airborne survey of Elliot Lake, Ontario. The readings were so intense at 500 feet that they couldn't be measured. I watched the equipment myself, and it looked like it was going to self-destruct. This huge tape with a red stylus on its spoked drive was just hammering away. It was unbelievable. We could not measure the intensity of the signal; it was completely off the scale. I had never seen anything like it. I'd almost bet that our aeroplane glowed every time we passed over the area.

On another occasion we were asked to take our Canso and look for an Austin Airways machine and crew (another Canso, used on transport work) that had gone missing. Working out of Great Whale, they were heading for their home base at Timmins, Ontario, and were just two miles out when both engines started to miss. The pilot didn't panic. He turned on his landing lights, crossed himself, and pulled back on the control column, dropping the aeroplane into the bush. He didn't even take out many trees.

We had just taken a direct line toward Great Whale from Timmins when Gord, my co-pilot, nudged me.

"Smitty, there's smoke there." He had seen a black smudge rising above the trees.

"Let's have a look," I said.

So we turned the aeroplane and spotted this red tail just showing above the treetops. Now, the old Canso needed a couple of miles in which to make a turn. So we picked a reciprocal heading and returned, coming in low, studying the tops of the trees. We could still pick up the smoke, and, when we dropped lower, we spotted the wreck and only one of the crew. We knew that there were two pilots on board, and the flight engineer—who was there waving at us.

When that 16-ton aeroplane went down among the trees, the nose was crushed and both pilots were thrown against the Canso's massive control column. Each suffered chest lacerations and broken ankles. We circled, pinpointing the location, then radioed that we could see one survivor. We suggested they get a helicopter up there and pull the other two guys out. The incident ended with their rescue, and we were glad to have helped. Incidentally, our employers gave us a week in Florida for our efforts.

As tough as the job was, I wouldn't have traded it for anything. I had come into it at 22 years of age with only a couple of hundred hours of flying time. I flew to a lot of places, met a lot of people, and saw a lot of things. It was fun. Everyone was good to me, and I feel I was very fortunate, at such a tender age, to have done the things that I did. The job was harder on the young married guys (as I was soon to be), but they got caught up once they reached home. Being engaged in exploration work made huge demands on everyone's time. But it was worth it. I have great memories.

"Smitty" and wife-to-be, Elinor, with one of their employer's Cansos.
Via E. Smith

Bush Flying Summers

When Long Hours of Daylight Made for a Rigorous Schedule

A. B. (ART) WAHLROTH

Art Wahlroth had a distinguished service career in the RAF and RCAF, which included two tours piloting Wellington bombers over Europe with 405 Squadron (RCAF) and in the Middle East with 37 squadrons, separated by a stint instructing on an Operational Training Unit. He returned to Canada in 1944. Flying out of Rockcliffe, Ontario, with 413 Photo Survey Squadron, he managed to acquire a few dozen hours on the three elderly Spitfires and the Hurricane that were part of the Rockcliffe complement. Of this flying time, he is especially proud. And it was during his RCAF service in Canada that he became acquainted with the Noorduyn Norseman and began a lifelong love affair.

Art left the RCAF in 1948 and decided to combine further education at the University of Toronto and later the Ontario College of Optometry with commercial flying in the summers. To this end he obtained a commercial pilot's licence at Toronto's Barker Field. His flying activities would provide the necessary funding for his schooling.

He spent his first summer operating a one-man flying school at Owen Sound, Ontario, and the next season barnstorming along the St. Lawrence River. In 1951, he flew a Republic Seabee for Georgian Bay Airways, and in 1952 joined Austin Airways operating out of their Sudbury, Ontario, base. In his three seasons with Austin, Art flew

Beaver, Norseman, Stinson 108, and Cessna 180 aircraft, all on floats. For the summers of 1956 and 1957 he moved farther west to fly with Ontario Central Airlines in northwestern Ontario. He spent his last full-time flying season with Wheeler Airlines, a long established firm based in St. Jovite, Quebec. Although he hung out his shingle as an optometrist, he continued to fly privately. Among the aircraft he subsequently flew, he most enjoyed a home-built "Breezy," an extremely basic aircraft—reminiscent of the earliest years of aviation—which the pilot sits on (rather than in) and is fully exposed to the elements.

His flying experiences differ in time, locale, and cargoes from those of Bert Phillips and Jeff Wyborn. While there are similarities that make for interesting comparisons, the differences are significant. Although Art flew a Norseman servicing construction crews on Mid-Canada (radar) Line construction, most of his flying was in the more northerly tourist regions, servicing hunting and fishing lodges. His sportsman passengers tended to be less predictable than the seasoned miners, trappers, prospectors, fur traders, and northern residents recalled by Bert and Jeff—who took flying in stride. His flying was also more hectic, with shorter but more frequent trips—and many more landings and takeoffs, when accidents most often occur.

It has been said that the best sea plane pilots were also capable sailors. On the water a floatplane—because of its bulk—is at the mercy of the wind, not unlike a sailboat sitting with its sail raised. Docking can be especially tricky. How best to manoeuvre using the prevailing wind, the water rudders on the floats, and bursts of engine power, combined with the aircraft's (air) rudder can only be learned from experience. Art Wahlroth provides insight into this important but often overlooked phase of northern flying. The reader may detect a certain feistiness—for which Art is well known among his friends. Art Wahlroth has spoken on several occasions to both the Society's Toronto and Ottawa chapters on different phases of his flying career. His bush flying account appeared in two successive instalments in the Spring and Summer 1996 *Journals*.

At college, when asked how I was going to spend my summer, I responded that I would be flying in the bush. Invariably I would be asked the same question, "Isn't that dangerous?" "No," I would reply, "the dangerous part is driving there and back."

My kind of bush flying was physically and mentally demanding, but if you knew what you were doing it was reasonably safe. I spent my 1955 and 1956 summer holidays with Ontario Central Airlines at Kenora, Ontario, where I had always wanted to fly because of the nature of the countryside. The previous summer I had flown out of Sudbury with Austin Airways. Kenora, Central's main base, was two days and 1,235 miles from Toronto by Volkswagen. A Native-sounding name with a nice ring, Kenora was actually a combination of the first two letters of three adjoining municipalities: Keewatin, Norman, and Rat Portage.

A Kenora ordinance would not allow us to take off in the inner harbour, so we were obliged to taxi quite a distance to keep the noise away from town. This will later have a bearing on one of my stories. The Lake of the Woods is so full of islands (promoters claim 60,000)

Ontario Central Airlines' base at Kenora, Ontario, in 1956. OCA aircraft are to the right of the dock. Distinctive gull-wings identify the Stinson Reliants. The other two aircraft are Norsemans. *A. B. Wahlroth*

that I had to fly above 1,000 feet in order to navigate. At this altitude, anyone—guide or lodge-owner—who knew the lake, became completely lost. Conversely, if I flew at 100 to 200 feet, my passengers knew every shore and island—and I was the one lost. Thirty years later I viewed the lake while a passenger in a Boeing 737 at 7,500 feet. It was the first time I had seen all of it at once. What a magnificent sight.

OF FISH AND FISHERMEN

Most of our flying from Kenora involved taking tourists and supplies to and from the fishing lodges on the Lake of the Woods or the English River system. The chief lodge operator, and a helluva nice guy, was Barney Lamb, who also controlled Ontario Central Airlines, monitoring it from his Ball Lake Lodge by radio. Don Watson, the general manager, ran the line from Kenora. Barney had a private home near Ball Lake Lodge, where he lived with his wife, four or five young daughters, and a gigantic Irish wolfhound. Ball Lake is part of the English River system. Unfortunately for those waters, a paper mill had for years been dumping mercury into Clay Lake, which feeds into Ball, then Grassy Narrows, and on into the Winnipeg River. Barney was one of the first people to draw attention to this menace and attempt to rectify it. But apparently the paper interests had greater clout with the government than did lodge owners and Indian bands. It was not until many years later, after the damage was done, that clean-up was even attempted.

Barney had sold Austin Airways a Norseman and taken a Cessna 180 in part payment. Since I had flown this particular aircraft in Sudbury, it was given to me. The 180 was a very versatile workhorse, and I could take two or three passengers into a small lake—rather than one at a time with the smaller Piper Cub. The Cessna did such a good job that first summer that when I returned the following year, Barney had purchased five new 180s and hired less-experienced pilots to fly them. I was relegated to freighting with a Norseman.

Float-equipped aircraft are licensed to operate under visual flight rules (VFR), from a half-hour before sunrise to a half-hour after sunset. In midsummer, in the North, this can be from 4:30 in the morning until 10:30 at night. Ontario Central Airlines was so busy at this peak season that we didn't have much time between trips. I might bring one machine into the dock at Kenora only to be hustled straight into

Perched on the float of his Ontario Central Airlines Norseman, author Art Wahlroth makes friends with an Ilford, Manitoba, resident in 1956. *A. B. Wahlroth*

another and sent on my way again. Don Watson would meet me at the dock and hand me my paycheque. It went into my pocket right next to the previous cheque, which I hadn't yet found time to bank. My log shows that one day I spent nearly 10 hours in the air, which may not sound like much. But remember that the average trip was about 50 miles—30 minutes flying time—and the aircraft had to be loaded and unloaded at each end of the trip.

Because of the demands of the job, I was always appreciative of small kindnesses. One of my warmest memories is of a cook at one of the lodges. Learning that I had not eaten that day, this lady cut me a slice of tender roast beef, still hot, and wrapped it in separate strips that I could eat in the air. It was delicious, but hardly a meal. Small wonder that I would lose 18 pounds in a summer.

One morning about 5 AM, as I walked across the parking lot to the local restaurant, I realized that I had gotten up, dressed, and left for work—without cleaning my teeth or shaving. Fatigue can cause one to operate at a pretty high level of stupidity. Fortunately, Barney Lamb realized this and would occasionally send us up to one of the cabins to catch an hour's sleep. Visiting Ball Lake first thing in the morning and taking on a load of gasoline or supplies was almost routine. After breakfast I might ferry a fisherman (with guide and canoe) into a nearby lake and then return a planeload of departing guests to Kenora. Since each passenger proudly sported a bag of fish, we would notify Kenora by radio, and the fish-processing people would be waiting for us at the dock. As the season progressed, I acquired a fair amount of fish myself, and, when the processor shipped them to me later in Toronto, he included some venison and moose meat.

Barney operated a northern camp at Kanuchuan Rapids, Ontario, a cabin at Elk Island on God's Lake in northern Manitoba, and another about 50 miles down God's River. Our routine, after taking a party to Kanuchuan and spending the night there, was to fly another party to God's Lake for lake trout fishing, and a third group from there to the river cabin, for speckled trout. Later I would return the original group to Kanuchuan and then Kenora. Our fisherman clients seldom went away disappointed. Lake trout in God's Lake went up to about 40 pounds, and the speckleds from the river were from three to six pounds.

One evening as I was returning from the river cabin with only one passenger on board my 180, I received instructions to land at God's Lake and pick up two more men who were fishing there. Landing beside their boat, I taxied into the lee of an island and loaded them on board. All three passengers were big men with lots of gear, and the latter two were soaking wet—with about 200 pounds of fish. This was a considerable load for a four-place aircraft. The Cessna's floats were awash.

I asked the guide if I could take off straight ahead from the island, and received an affirmative answer. But when we got up, running on the step at near takeoff speed, I noticed ridges of rock sticking out of the water ahead. They were fast approaching, and there were only minimal gaps between them. But I was committed—and could not stop. How I managed to weave through I don't know, but we made it. That guide's ears burned for the rest of the night.

ODDBALL PROSPECTORS AND MEMORABLE PASSENGERS

The big mining interest at the time was the uranium that cropped up in odd places along the edge of the Laurentian Shield, the western edge of which roughly follows the boundary between Ontario and Manitoba. I had occasion to take two men, one looking like a woodsman, the other a pale-skinned city dweller, up to a point where the north-south line of the said border turns northwest towards Hudson Bay. The pilot that sets people down in the bush always returns to get them. When I flew in to collect these two a few days later, they were the most miserable, dirty, wet, tired, mosquito-bitten pair ever cheered by the cold beer that I always carried in the floats for just such occasions. Apparently they had been trying to reach a nearby lake— too small to get into by air—to obtain uranium findings; but the bush was so thick they failed to get through. I learned from them that the "city dweller" was a hypnotist, who discovered that some people, under hypnosis, were sensitive to certain minerals and could locate them on a map. They were trying to prove his theory. Some time later, I noticed that a survey map from a seismographic study did show uranium at just that location.

One of the camps we serviced on the English River was Keith Hook's Separation Lake Lodge. On one trip in I had a Norseman load of 45-gallon drums of gasoline—and one hung-over guide who smelled so bad I wouldn't have him up front with me. After frisking him for matches and cigarettes, the dock crew loaded him in the back with the gas. We were nearly over Separation Lake when I detected the sulphurous smell of burning matches! I tossed the machine around a bit to keep my passenger off balance and quickly set it down on the lake, by now under me. When we stopped I jumped down on the float, opened the back door and hauled him out, right into the water, which was cold enough to sober him up. But I was still angry. I wouldn't let him crawl back on the float. Instead I towed him across the lake to the camp. Did he realize the catastrophe he could have caused? I wonder.

Part of my time with OCA I spent at their Red Lake base. This gold-mining town was built on rock, and at that time had no water system. When you wanted a bath you had to go to the local hotel and pay 50 cents.

Art Wahlroth checks an aileron on his Norseman, CF-CRC, at Keith Hook's Separation Lake Lodge on the English River system in northwestern Ontario. *A. B. Wahlroth*

I met all sorts of people there. On two successive trips I dealt with contrasting types, first the thoughtless and inept and then the very sensible. My first trip was to pick up a party of campers from an island. There was a brisk wind that day, and the dock on the island was in such a position that I needed help to get in, particularly as there were two small boats tethered at the end. My passengers-to-be sat there and watched me taxi in circles making a half-dozen futile stabs at reaching the dock. Either they did not appreciate my plight—or didn't care. I suspect the latter. Finally, I shut off the engine and hollered at them to clear the boats away and catch the wing strut. At long last they came to life and helped me to dock.

One of their group was an elderly man, whom I put in the front seat and handed an insect spray bomb to use *if necessary* while we were loading. When I climbed back into the aircraft, I found that he had completely obliterated the windscreen and both side windows with a thick film of spray! I carried no rags in the machine; before we could get under way they had to unpack a suitcase and clean the windows with their underwear.

My next trip was in gratifying contrast. I took a couple of men into a lake where I had to back the aircraft onto a beach. Without saying a word, they took off their shoes and socks and rolled up their pant legs. Then each climbed out on a float and dropped into the shallow water to guide the machine back onto shore.

ROUTINE HAZARDS: LEARNING FROM EXPERIENCE

Perhaps the closest I ever came to sinking into the bush came at Red Lake. There was a large power line between the mainland north of the town and Campbell Island; if you took off to the west you started your run from that point. To the east, however, there was enough water to get off—if you were lightly loaded. But if you were heavy you had to taxi back under the wires and, making your run, hold the machine on the water until you got under the wires. Then you gunned it to get off. On this particular day, my load seemingly consisted of several sheets of plywood, along with two passengers and their suitcases. Blithely I started my takeoff run to the east from the wires. I got the machine off, but only just! Had there not been a river to follow, I'd probably be there yet. At my destination, I removed the plywood and found a load of cement in bags, which the ground crew had forgotten to mention to me. I should have taken a closer look at the water line on the floats.

One day, flying my 180, I took a couple of passengers from Ball Lake, Ontario, to Kenora, and then on to Winnipeg. In our path was a line of cumulonimbus clouds that were hitting down to the north and south, but with a clear patch in the centre through which I could see the sun shining on the Manitoba wheat fields. It was a typical line-squall heading east, with an equally typical roll-cloud at its leading edge. I explained to my passengers that it would be a bit rough going through, but we could do it. They had been drinking their lunch at Kenora's Kenricia Hotel, so they were feeling no pain. We passed through without mishap. But I had to dive when going under the roll-cloud to avoid being sucked up into the thing and to climb when at the back end. Two Norseman aircraft were less fortunate. They landed at Separation Lake and were caught on the water. It would have been worse had they been in the air. Hailstones as big as your fist punched holes all over those aircraft, and both machines had to be recovered with new fabric before being flown out.

The upper rear fuselage of an Ontario Central Airlines Norseman caught in a hailstorm. Had the aircraft been flying the results would have been disastrous. New fabric had to be applied on the spot.
A. B. Wahlroth

One of the more elaborate wilderness docks built entirely with spruce poles on the shore of Pot Hole Lake, Manitoba. The Bombardier tracked vehicle in the background was the only means of surface travel over the surrounding muskeg. *A. B. Wahlroth*

That particular summer I spent a few weeks on the Mid-Canada radar line, flying a Norseman out of Ilford, Manitoba. There were three radar lines: the DEW line in the far north across Baffin Island, the Pinetree line, which crossed over about the Sudbury-Kenora level, and the Mid-Canada line in between, crossing over the Hudson Bay flats. The Bell Telephone Company was the contractor, and our job was supplying the Doppler radar sites, which were being built on hard-ground strips that were remnants of ancient beaches. The whole country was muskeg; you could not walk across it in summer. From the air you could see the winter roads stretching from lake to lake where supplies were brought in by tractor train over the ice and snow.

At each site, we had to land on the nearest lake, usually a round, shallow pond, only deep enough for us to set down in the exact centre. As we taxied toward shore our floats would bump onto rocks, which would be pushed down into the silty bottom by the weight of the machine. To take off we taxied to the centre of the pond, shut off, and were blown back till the floats grounded. Mushy though it was, you could touch firm bottom with a paddle, anywhere.

195

This country boasted "bulldogs"—gran'daddies of the horsefly—about the size of lean bumblebees. These creatures would hit you, take out a chunk, and keep right on going. On approach, say within 200 feet of the water, we made sure the windows were closed, that head-net and gloves were on, and trousers tucked into boot tops. We could hear the insects drumming on the fuselage of the machine. As soon as we had unloaded we closed up the aircraft and sprayed the interior to keep it habitable. It wasn't unusual, after three or four trips, to sweep out a bucketful of fly corpses.

When I left Kenora for northern Manitoba, the flying instruments in my Norseman wouldn't work. The mechanic accompanying me assured me he would replace them at Red Lake. However, Red Lake didn't have any—they would have them sent up. But they never arrived. I spent several weeks with few instruments until, finally, not even my air speed indicator functioned. So it was that, getting caught in the smoke from bush fires, I stayed on a lake for three or four days. Eventually I flew out through the smoke, formating on a Canso,

Early morning at Wheeler Airlines' St. Jovite, Quebec, base. A DHC-2 Beaver (foreground) and a Norseman ride at their moorings.
A. B. Wahlroth

which kept its speed down to mine until we emerged from the smoke, then he ran away from me.

At the end of the school year at Red Lake, we had to take the Indian children back to their reservation at Big Sandy Lake—with special Department of Transport permission to carry more than 10 passengers in a Norseman. While our two machines were being gassed up, a boy in a small boat started making circles out on the water, creating waves that rocked the aircraft. This made it difficult for the men up on the wing doing the refuelling, much to the amusement of the children and, indeed, the entire population of Red Lake, which had turned out to see them off. At the critical moment in a turn, his motor stalled, rocking the craft and throwing him into the water. To hilarious laughter, he dog-paddled back to his boat, probably red-faced. Soon I was able to take off with 18 little kids packed into the machine. Things went well until one of them got sick—then they *all* got sick!

One night, before finishing up the season at Kenora, I was asked to make an emergency trip to a mine 50 miles away, at dusk. The half-tank of gas in my machine was enough. I wasted no time taking off in the inner harbour—the prohibited area—directly over a largish boat that turned out to be the RCMP patrol craft. Though it was pitch dark, finding the mine was no problem. It was the only place with an outdoor electric light, hung from a pole. I got down, taxied up to the dock, and three men with toolboxes boarded, one of them bleeding badly. I put the two uninjured men off—taking off at night was too risky with a load—taxied into a bay, shut off, and floated back until the tail touched the bush. This gave me ample room to take off. When we arrived back at Kenora, the RCMP and Ontario Provincial Police boats were clearing the lake of craft in the area where I intended to land. I taxied in on the step, escorted by both boats. The RCMP didn't say a word about my taking off over their heads.

Driving home through the States I stopped at Bemigi, Wisconsin, where a flying outfit was hopping passengers. There is a huge statue of Paul Bunyan and his giant blue ox, Babe, here. This amount of bull must have had an effect on the inhabitants—one anyway. When a woman asked me if I would like an aeroplane ride, I told her that I had just returned from flying 300 hours during the past summer. "Oh," she replied, "each of our pilots does that much in a month!"

COPING WITH MORE RUGGED GEOGRAPHY

I spent my final summer of commercial bush flying with Wheeler Airlines at St. Jovite, Quebec, and again the aircraft were the Norseman, Beaver, and Cessna 180. Whereas in northwestern Ontario and Manitoba the hills are mostly low and uniform, with decent-sized lakes, in Quebec the real estate is rough and vertical—with noticeably smaller lakes. There were a few that I wouldn't go into until someone came along with me to show me how to get out. Generally speaking, we went over the crest of the last hill, throttled back, put on full flaps, and started to sideslip. But before doing this we circled a few times to confirm that there was a way out. Taking off was always a compromise between the wind, the longest stretch of the lake, and the lowest ground at the edge. One north-south lake in the area seemed to act like a tidal basin. There were two similar docks on this lake, one at either end; depending on the wind direction one dock would be awash and the other one three feet out of the water. With a wind switch this would be reversed.

Lac Ouimet, where our base was located, is the smallest lake from which I have ever flown a fully loaded Norseman. There were a couple of islands in the centre of the lake—if you weren't off the water by the time you got to them you didn't go. On one occasion I had on board two passengers, their gear, and a huge St. Bernard dog. As we taxied out they were all looking over my shoulder. On takeoff I could not get that machine up on the step. Looking around, I discovered that as soon as I started my run the St. Bernard slid to the back of the cabin, and there was no way I could lift that 250-pound dog up on the step! When they held the dog forward there was no further trouble.

Language presented a bit of a problem for me, since all communication was in French, including the company ground-to-air radio. But just about everyone had a smattering of English and was eager to practise (a bilingual observer on that frequency would have had a lot of laughs). With my high school French I got by—almost always. But once, after I had taken some passengers to the Chapleau Fishing Club, my aircraft was being held to the dock by a couple of guides. I asked them, or so I thought, to let the nose swing free and to hold the tail. But just the reverse happened, and I had to reach out to the chap who had just released the back of the float. "Bonjour, monsieur," he greeted me and gravely shook my hand.

The most spectacular real estate in the area was Mont Tremblant, a famous ski resort not far from St. Jovite. When I was returning to base, coming over the top of this mountain, I would throttle back and—keeping clear by about 100 feet—would wind back and forth down the slopes, following the ski runs. I became rather well known for this, and had people asking me for rides. I even took the airline bosses and their friends up to experience it. One, a pilot himself, said it was the biggest kick he had ever gotten out of flying.

A trip that every Wheeler pilot looked forward to—in fact, we operated on a rota—was collecting a particular customer and his family from Montreal and taking them to their summer place, an abandoned lumber camp 100 miles north. This chap would specify that we be at his dock on Montreal's Back River at a certain time—not five minutes early or five minutes late. He would have everything organized, luggage and family, which he would help load. Then he would climb into the front seat and map-read his way up. At the end of the trip he would invariably hand the pilot $50, a very nice sum in those days. Once he was a bit short, and apologized for only having $45.

HARD WORK AND TRICKY LANDINGS

On another less desirable assignment I had to move building materials from one lake to another up near La Tuque. I started out early, over some lifting fog, with the aircraft full of fuel and even extra containers on board. Landing on a lake north of La Tuque, some 100 miles north of St. Jovite, we found the dock jammed full of building materials and furniture—including even the proverbial kitchen sink! This was all to be moved to a half-built camp on a lake about 10 miles away, and it was quite a trick to do it. It took me all day and 18 separate trips, and if you think that wasn't work … !

There was a stiff wind from the west. In order to take off I had to taxi to the eastern end of the lake down a long, narrow gut, where I had to shut off the engine and let the aircraft swing around into the wind. For every takeoff the aircraft was at gross weight—as my fuel was consumed, the load increased.

Getting into, and out of, the other lake, was just as tricky. I had to land into the wind in a cut between two hills, run across smooth water protected from the wind by a high hill, and turn downwind before I got caught in the wind again. Should I be caught I had to "sail" the aircraft back to the dock, which would take about 15 minutes. Getting out was something else. There wasn't a long enough stretch into the wind to get off, so I had to get the aircraft on the step into the wind then turn down the lake, take off crosswind, and swing over to the lee side of the hill. When operating in Quebec, it was vital to read the country.

During the final weeks of my stay at Lac Ouimet, I spent some time restocking the headwaters of the river systems of the lower part of the province, dropping fingerlings of various species from the fish hatchery at St. Faustin, near St. Jovite. The fish people installed a water tank in the Norseman, about seven feet long, 15 inches wide, and three feet high, with a pipe running to a large funnel protruding from the floor of the machine at the back, stopped with a removable plug. Having decided what fish they wanted in what lake—usually a pothole on top of a mountain—they would map it all out. Then they would bring the fish over in wire baskets and put them in the tank, half-filled with St. Faustin water.

Once we had located the dropping point and decided how to get into and away from it, a man riding with me would run some water

A Norseman and a Cessna at Wheeler Airlines' dock provide the background for a scene from *The Plouffe Family*, a popular 1950s Canadian sitcom set in Quebec. *A. B. Wahlroth*

and put one or more baskets of fish in the funnel. I would approach the lake, letting down to 100 feet or less; he would be ready to pull the plug and drop the fish. Sometimes there were government men in canoes on the little lake observing the drops; they reported that 90 percent of the fingerlings survived. We ranged from some lakes near Ottawa right up to La Verendrye Park.

One of Wheeler's good customers had a house on a lake near Joliette, and a hunting lodge farther north where he spent weekends with his friends. I had ferried him back and forth several times, and he took great pains to point out the location of all the power lines on his lake (Joliette). He missed one. Though I had seen sailboats in this particular area of the lake and looked the place over thoroughly, I managed to snag a low-hanging wire as I came in to land—out near the middle of the lake! The good old Norseman hardly felt it. The wire snapped, wrapping around the spinner and a wing strut, and tore some fabric off one wing. Safely down, we peeled off the wire and sealed the wing with some adhesive tape from the local pharmacy. This was the only air accident I ever had. The Shawinigan Power Company billed the airline $175 for the wire.

On my final trip for Wheeler—I was going back to school, and they had already hired a replacement for me—the weather turned sour, and I was given the option of finishing out the day or letting my replacement start in. Not liking the weather, I opted to quit immediately, which was a good choice. The new chap, being unfamiliar with the country, was forced down, and had to spend a couple of days in the bush.

The foregoing is now history—I wish it weren't—I'd love to do it all over again.

Water Bombing:
A Demanding Job

Flying Low and Slow
through Smoke and Ash

ROBERT S. (BOB) GRANT

Bob Grant's story differs from the other accounts in this book in that it deals with a contemporary aspect of bush flying using a relatively modern aircraft. The other stories relate to bush flying from the early 1920s to the 1960s. Bob's story, which describes a situation 30 years later, makes for fascinating comparisons.

The Canadair CL-215, for instance, with its maximum loaded weight of almost 22 tons and span of 93 feet, has a slightly lesser wing spread than its Canso predecessor in the water bombing role; but is nearly 25 percent heavier and has almost double the power.

And if one compares the modern CL-215 to the HS-2L from the very earliest years of bush flying, the differences make very apparent the incredible advances in aviation over just a half-century. While both are flying boats, that is where the similarity ends. Fully loaded, the all-metal CL-215 weighs almost seven times as much as the wood-and-fabric HS-2L at maximum weight, and, where the vintage Curtiss could carry the contents of a 45-gallon drum only with difficulty, the 215 can load six tons of water in seconds. And, not surprising, the Canadair machine has 12 times the horsepower. The complexity of Grant's pre-flight checkup also contrasts with the simplicity of such a process in earlier years. But, where it was sometimes necessary for those aboard a loaded HS-2L to bounce up and down in unison to induce their aircraft to leave the water during a takeoff, the CL-215 literally leaps into the air. But there is one similarity: the strenuous effort required of the crew remains unchanged.

Bob Grant is not only a very experienced bush pilot but is an established aviation writer with two books and innumerable articles under his belt. A native of Belleville, Ontario, he makes his home in North Bay, Ontario, where he flies CL-215s for the Government of Ontario for half of the year and spends his remaining time writing. His CL-215 story, entitled "The Slab-Sided Yellow Barn," was first published in the Fall 1992 issue of the CAHS *Journal*.

To pass the time before someone authorizes our departure from this God-forsaken (but unnamed) town slapped down on the edge of a boreal swamp, I dig out pad and pencil. I might as well write as I wait for a "Red Alert," which means snap the aeroplane into the air quickly—someone has discovered a forest fire in the rain-soaked spruce surrounding our airport. I have plenty of time.

THE CANADAIR CL-215 AND ITS PROCEDURES

Canada has 48 operational Canadair CL-215s. In the minds of those who fly the "slab-sided yellow barn" as some have nicknamed it, the CL-215 stands out as the greatest water bomber of all time. My employer (the Ontario Government) operates nine such machines. Some are theirs, and the rest are "borrowed" from the federal government. Few fly more than 200 hours annually.

Our tanker weighs 27,529 pounds empty, without fuel. Fuel cells between the engines can hold 9,360 pounds, but most pilots carry 5,000 pounds of 100/130 octane gasoline and 450 pounds of oil for standard fire-fighting work. Fully loaded with 12,000 pounds of water, we can legally leave a lake at 43,500 pounds gross weight. The airplane averages up to 1,000 pounds per hour fuel burn. In cruise, leaning the engines brings it as low as 752 pph at 5,000 feet.

The twin row, radial 2,100 hp Pratt & Whitney R-2800 CA-3s have 36 spark plugs each. Like most heavy piston powerplants, these engines demand considerable maintenance. They were designed for airliner use—long climbs to altitude and a cruise to some distant destination like Hawaii. No one intended them for use in the on/off stresses of fire fighting. Maximum-power takeoffs followed by idled

Canadair CL-215 C-FAFO is owned by the Government of Canada but operated by the Province of Saskatchewan. *R. Grant*

throttles minutes later, hour after hour, means frequent "jug" (piston and cylinder) changes, spark-plug changes, and even engine changes.

In the field, our organization depends on a CL-215 checklist, and to date, not one aircraft has been involved in a serious incident. Many items on the list resemble those a Cessna 172 or Piper Cherokee pilot would expect, while others are typical only for heavy aeroplanes.

In our case, one pilot performs a "top check"—he throws a protective pad on the copilot's seat and climbs through a hatch for a visual inspection on the one-piece, fail-safe 16,900-pound constant-chord wing. The aeroplanes's tail is a daunting 29 feet, nine inches high, and although the wing may be lower by a few feet, the asphalt below is only a horrifying slip away.

On top, the hero—usually a copilot—leans forward on the wing leading edge to press four plungers with either a calloused thumb or screwdriver to drain fuel tank condensation. Engine oil must be checked by unfastening a *dzus fastener*, unscrewing a cap the size of a pie plate, pulling out the dipstick and wiping it on one's arm to ensure that at least 20 imperial gallons exist. Consumption of two gallons per hour is not unusual on some CL-215s.

Behind the ash-, oil-, and carbon-covered right engine nacelle, a

Briggs and Stratton ground power unit, or GPU, weighing 185 pounds needs an oil check as well. When we stop on a lake—something rarely done with a CL-215—the GPU's weight helps keep the right float down and the left doorsill clear; otherwise, the cabin would fill and sink the aircraft. Unlike the Canso, CL-215 tip floats are never in the water simultaneously nor are they retractable.

Back inside, someone starts the GPU and drags the seven-rung boarding ladder inside and smashes his head while swinging it around towards the tail. There, it will be fastened on the right wall with two nylon straps. At this point, particularly when rushed on a fire dispatch, pilots would sometimes like to invite all assembly line workers aboard and have them click the silver fastener buttons in place. Someone deep inside the bowels of Canadair sometimes neglects to make the straps long enough. Consequently, a rushed crew ties the ladder with a rope.

With GPU generator on and its throttle set at "run," the CL-215 is now ready for the *Start Right Engine* portion of the checklist. Someone—usually the captain—calls for 15 blades on a warm start; 10 for a cold one. At the end of the count, the pilot switches magnetos to "both" and primes with a small toggle on the overhead panel.

The 14-foot, three-inch, three-blade Hamilton Standard propeller rotates slowly, and, after what seems a very long time, a "thud-thud-thud" and oil pressure rise indicate a successful start. Bluish white smoke erupts from the cowlings and dissipates quickly into the slipstream as the non-starting pilot eases the mixture control forward.

Occasionally, black smoke doodling from under the cowl flaps calls for another mixture adjustment. The left engine can be started almost immediately since the GPU does not require charging, unlike the batteries of some turbines. In any case, idle must be kept below 1,000 rpm until minimum oil temperatures show on the gauges.

The CL-215's *raison d'être* centres around a pair of water tanks located almost on the aircraft's centre of gravity. The lower two-thirds below floor level form part of the hull, and a fiberglass section makes up the portion visible from the cockpit. With bomb system master on—two oversize toggle switches behind the power quadrant—both pilots see five green lights on the hood of the instrument panel.

Dead centre on the top panel, an arm switch glows orange when activated, and a touch of a button on the left or right control wheel

Owned and operated by the Government of Ontario, tanker 268,
C-FDFR is seen after a rare precautionary water landing on a lake south
of the Sudbury airport in 1991. The immense size of the vertical tailplane
is apparent. *R. Grant*

opens the doors. In emergencies, either pilot can yank a red handle to
mechanically dump the load.

With both engines running, more steps on the checklist bring the
aircraft to the "pre-takeoff check." In position on the runway, five
items must be completed before the roll, including setting transpon-
der code 1276 if the airplane has been dispatched to waterbomb. Both
crewmembers ensure runway headings match the compass. After centring the hydraulically steered nosewheel, the left-side pilot releases a
control lock on the power quadrant. It cannot be overlooked since the
throttles will not go past a detent.

Aviation enthusiasts watching Canadair CL-215s leaving airport
runways often gain the impression that the big yellow birds are nothing more than overweight pigs staggering into the air. Inside the
cockpit, however, things happen fast. As the flying pilot—in our
group, pilots share responsibilities but do not switch seats—brings
power to a maximum of 53.5 inches of manifold pressure; directional
control requires concentrated effort via a tiny hand-operated wheel
below the left side window.

The rudder takes effect at 30 to 40 knots, and the massive CL-215 gains momentum towards its V1 of 86 knots. As the gigantic collection of Plexiglas, metal, rubber, plastic, and human beings thunders down the runway centre line, both pilots feel every shattering explosion in each of the 36 cylinders.

As speed, shaking, and noise increase, every cable and wire in the cockpit vibrates itself into a blur. On hot summer days, we welcome air blasting from the vents above our foreheads, but the shrill piercing sound adds to the din. Frequently, aeronautical charts, checklists, or approach plates dance out of the holders beside our seats. On every trip, we thank God and the almighty taxpayer for the Clarks (headsets), without which our hearing would never survive.

At 91 knots, a slight backward movement of the control wheel rotates the aeroplane, and the flying pilot calls for gear up. His teammate reaches across and flips a three-inch plastic wheel located a little left of the engine instruments. As each black rubber tire settles into place, a call for "flaps up" comes next. At 500 feet above ground, we reduce power to 45 inches and 2,600 rpm, and soon another call follows for 35 inches and 2,400 rpm.

With the main job of getting into the air completed, the non-flying pilot carries out "post-takeoff checks" once traffic warrants enough time to bring his eyes back into the cockpit. A quick glance at the engines ensures that pieces of cowling have not blown away.

Most pilots cruise the Canadair CL-215 well below 10,000 feet at speeds averaging 138 knots. Cylinder head temperatures must read between 190 and 200 (C) and are kept in range with cowl flaps. If time and distance permit, the mixtures can be pulled back to lessen fuel consumption. Carburettor heat rarely becomes necessary on the CL-215.

WATER BOMBING

Enroute to the fire, navigation is done with the aid of LORAN (long-range aid to navigation) backed up with map reading. Ten minutes back from a fire, the crew calls the "bird dog"—usually a Cessna 337 or Aero Commander—with altitude and position information.

Using a non-airconditioned, multi-ton, slab-sided, flying fire truck to drop 12,000 pounds of liquid into smoke, turbulence, and rising terrain is not a matter taken lightly. Few outsiders realize the

CL-215 lacks boosted (power-assisted) controls, and that ailerons require sheer muscle power. At 180 pounds, I am not considered physically strong, and crosswind water pickups require every ounce of strength I can find from the tips of my fingers to the muscles that flex my toenails.

On good days, water work requires one hand on the control wheel as the other manipulates the throttles, but in crosswinds, I cannot hold the wheel into wind with one hand. After running engines to takeoff power after touchdown, I need both hands on the wheel. No question—a Canadair CL-215 is a two-person aeroplane.

Before the first pickup, one crewmember leaves his seat and strolls to the aeroplane's midpoint—the longest walk in the world on a turbulent day. He kneels down, face to the floor, and peers through a tiny plastic window to ensure that the landing gear springs and uplocks are in place.

Next, he checks the nosewheel doors by lying prone and peering through another Plexiglas plate between the pilots' seats. We have been told that some organizations operating CL-215s trust the warning lights implicitly. Ours does not, and considers the visual check a cost-free precaution.

Prior to water pickup, a switch near the landing gear handle must be flicked to "sea," otherwise the warning horn would blare every time the throttles came back. Also, pilots ensure that two red spigots indicating doors closed are visible forward of the water tanks and that another two in the well between the seats indicate nosewheel doors locked. A switch behind the mixture control levers lowers two, square, stovepipe-size hydraulic probes through which the water will rush into the tanks.

On approach, power should be reduced early in stages to allow cylinder heads to cool slowly. With flaps at 15 degrees, the aircraft crosses the shoreline at 90 knots. When the aircraft slams into the water at about 75 knots, a very distinct nose-down pitch shocks us every time. The flying pilot eases in the power and within six seconds of shuddering, shaking, banging, and roaring of both engines at takeoff power, 1,200 gallons of water smashes into the probes, runs up a pipe, and explodes inside the tanks.

Flick the probes up—the aircraft accelerates—and in another six to eight seconds it leaves the water at 78 knots. Flaps to 10 degrees,

The very basic angularity of the CL-215 is seen to advantage in this picture of an unidentified machine based in Yellowknife, Northwest Territories. *R. Grant*

wait until 95 knots and positive climb rate before arming the drop doors. Once safe at 105 knots, power reductions follow.

Arming too soon could be extremely dangerous. If our doors opened with the aircraft on pickup, no doubt the ensuing ride would be an exciting one, probably to the bottom of the lake. No one in the business intends to find out.

Our manuals show maximum drop speeds of 129 knots, but most "drivers" try for 110 knots. Some pilots ask for an additional five degrees flap during the drop. A few prefer having the non-flying pilot handle throttles while they keep both hands on the control wheel. A 12,000-pound load of water can do plenty of damage if dropped low enough. Usually, the flying pilot presses the button 100 feet above ground, but judging height can be difficult. Water bombing technology is *eyeball*—no laser-guided sights to place our loads through an Iraqi's kitchen window.

When the load goes, the slam of the 63-inch x 32-inch hull doors can be heard above the noise of the R-2800s. Water exits in little more than a second and the doors close hydraulically. Pre-pickup checks are carried out in an organized manner again, and we return for another pickup.

Picking up, dropping, picking up, dropping sometimes goes on for hours, but we are limited to eight hours fire fighting per day. On windy days, with greenish white splotches of smashed insects obscuring our vision and ashes sifting through the air vents, the job seems like a tough one. Both pilots work hard, although some observers think copilots do nothing except the logbooks. They will never know—passengers are not allowed on fire-bombing flights.

Once released from the fire, the aircraft climbs to a homeward altitude and someone completes post-takeoff and cruise checks. Usually at this stage, the paperwork gets finished—fire numbers, loads dropped, weights, cost codes, etc. Of course, a weight and balance must be calculated before landing or the wings may fall off. (As with most large aircraft, the CL-215 cannot be landed with too great a weight of fuel on board, especially as it is located in the wing tanks.)

Landing a CL-215 on a runway in light wind conditions is not difficult. Full flap of 25 degrees almost always comes into use even in the strongest crosswinds. The acting copilot ensures the gear is down by three striped indicators on the instrument panel. Approach takes place at 110 knots and touchdown at 75 to 80 knots. On the surface, a control lock must be engaged immediately by centring the column and using a pair of decalled white marks on the floor as a guide. The aircraft should never be turned downwind with control locks in place or elevator, rudder, and aileron stops could be severely damaged.

Cowl flaps open for cooling, electrics, auto-feather off, lights off, etc., are part of the post-landing litany. Before moving into the parking area, the crew drops the water doors to release residual water—mechanics rarely appreciate wet ramps. The doors remain open when the aircraft parks to prevent damage to the rubber seals.

After bashing his head on the ladder again, one pilot steps outside and inserts a steel pin into each main landing gear and a "horse collar" in the nosewheel well. These costly gadgets ensure the aircraft does not wind up on its belly if the hydraulic system malfunctions.

Back to "Red Alert."

A Brief Chronology
of Bush Flying in Canada

Much of the following data has been extracted (unabashedly) from *125 Years of Canadian Aeronautics: A Chronology, 1840–1965* by George A. Fuller, John A. Griffin, and Kenneth M. Molson. Edited by Fred Hotson, who was then president of the Canadian Aviation Historical Society, this excellent book was published by the CAHS in 1983.

5-8 June 1919

Pilot Stuart Graham and Engineer Bill Kahre collect an ex-U.S. Navy Curtiss HS-2L flying boat from the U.S.N. Base at Halifax and ferry it to Lac-à-La-Tortue, Quebec, for use by the St. Maurice Forestry Protective Association. Registered as G-CAAC, this machine is regarded as Canada's first bushplane, with Graham and Kahre the first bush pilot and air engineer respectively.

August 1919

Using radio-equipped Curtiss JN-4 (Can) aircraft, the Owens Expedition carries out the first aerial timber survey in what would later become Canadian territory. These flights mark the first use of radio in civil aircraft and the first flights into Labrador.

30 August 1919

W. R. ("Wop") May flies a police officer from Edmonton to Coalbranch, Alberta, in a Curtiss JN-4 (Can) in pursuit of a murderer, the first instance of an aircraft being used for law enforcement in Canada.

October 1919

The first photographic survey operation in Canada is carried out by the team of S. Graham (pilot) and W. Kahre (engineer).

17 January 1920

Air regulations are published in the *Canadian Gazette* and become law.

19 April 1920

The Air Board is constituted by Order-in-Council to regulate aviation in Canada.

18 June 1920

The first aircraft ever to arrive in Whitehorse is a U.S. Army Air Service de Havilland D.H. 4 to make arrangements for a proposed American round-the-world flight.

17 August 1920

W. Roy Maxwell (pilot) and George A. Doan (engineer), flying from Remi Lake, northern Ontario, become the first to reach James Bay by air when they land at Moose Factory in an HS-2L.

28 August 1920

W. R. May and G. W. Gorman fly nonstop from Edmonton to Peace River Crossing, Alberta, in a Curtiss JN-4 (Can).

2 September 1920

W. R. Maxwell carries 100 pounds of mail from Remi Lake to Moose Factory, Ontario, the first volume delivery of airmail in Canada.

24 March 1921

Sponsored by Imperial Oil Ltd., G. W. Gorman and E. G. Fullerton commence exploratory flights down the Mackenzie River from Peace River Crossing, Alberta, using Junkers JL-6s (F-13s) G-CADP and G-CADQ.

5 April 1921

E. G. Fullerton successfully tests the first of two homemade propellers on one of their Junkers JL-6s at Fort Simpson, Northwest Territories.

5 February 1922

J. Hervé St. Martin and W. R. Maxwell make the first winter flight into James Bay, using the Avro 504K, G-CAAE.

August 1923

Fairchild Aerial Surveys (of Canada) Ltd. begins flying operations with a single Curtiss flying boat.

November 1923

Fairchild Aerial Surveys (of Canada) Ltd. rents a Standard J-1, becoming the first Canadian company to operate year-round in the bush flying field.

22 April 1924

The Ontario Provincial Air Service (OPAS) accepts delivery of the first of 13 Curtiss HS-2Ls to begin flying operations.

23 May 1924

Laurentide Air Service begins the first scheduled air service in Canada, between Angliers, Lac Fortune, and Rouyn, all in Quebec.

3 September 1924

Laurentide Air Service operates the first regular Canadian Airmail Service, between Haileybury, Ontario, and Angliers and Rouyn, Quebec.

3 November 1924

C. S. Caldwell carries the first aerial stowaway in Canada.

4 November 1924

W. N. Plenderleith test-flies the prototype Canadian Vickers Vedette at Montreal.

10 February 1925

D. R. MacLaren forms Pacific Airways, taking over fisheries patrol from the RCAF.

October 1925

The Ontario Provincial Air Service uses HS-2Ls to fly men and supplies in to develop the Howey gold find at Red Lake, Ontario.

March 1926

H. A. ("Doc") Oaks and others form the Patricia Airways & Exploration Co. to provide an air service for the Red Lake district, beginning operations with a Curtiss Lark.

Spring 1926

Jack V. Elliot flies prospectors and miners into Red Lake, Ontario, using a Curtiss JN-4 (Can) equipped with skis.

June 1926

Fairchild Air Transport Ltd. is formed to carry out bush flying operations.

10 June 1926

Western Canada Airways Ltd. (WCA) registers the first Fokker Universal, G-CAFU, to be owned and operated in Canada.

July 1926

La Compagnie Aerienne Franco-Canadienne is incorporated in Quebec mainly to carry out forestry and photographic work for the provincial government.

11 September 1926

J .D. McKee and Squadron Leader A. E. Godfrey fly the first Canadian transcontinental seaplane flight, using a Douglas MO-2BS.

10 December 1926

J. A. Richardson incorporates Western Canada Airways Ltd.

27 December 1926

Western Canada Airways Ltd. begins operations from Hudson, Ontario, into the Red Lake district with H. A. Oaks flying the Fokker Universal, G-CAFU.

February 1927

Fairchild Aviation Ltd. is formed, absorbing Fairchild Aerial Surveys and Fairchild Air Transport Ltd.

22 March 1927

Western Canada Airways Ltd. undertakes the largest air-freighting contract to date in Canada, moving 17,894 pounds of supplies from Cache Lake to Churchill, Manitoba, by 17 April.

1 June 1927

Western Canada Airways Ltd. inaugurates a weekly service from Winnipeg to Long Lake, Manitoba, via Lac du Bonnet.

18 June–27 July 1927

Flying Officer C. L. Bath carries out first forest-dusting operations in Canada using a Keystone Puffer.

14 July 1927

Fairchild Aerial Surveys of Canada imports the first Fairchild FC-2, G-CAGC, into Canada.

17 July 1927

The Hudson Straits Expedition sails from Halifax to gather information on ice conditions using Fokker Universal aircraft.

27 July 1927

Squadron Leader T. A. Lawrence and Flight Lieutenant A. A. Leitch make the first flight by a Canadian aircraft in the high Arctic using a de Havilland D.H. 60 Moth seaplane, G-CAHK.

1927

The first McKee Trans-Canada Trophy is awarded to H.A. ("Doc") Oaks for his organizing of air transport in northern Ontario, Manitoba, and Saskatchewan.

13 February 1928

Prospectors Airways is formed for the purposes of aerial mineral exploration.

March 1928

J. E. Hammell and H. A. Oaks form Northern Aerial Minerals Exploration Ltd. (NAME) to employ aircraft in the search for minerals in the North.

8 March 1928

Roy Brown forms General Airways Ltd. at Toronto to operate from a base at Rouyn, Quebec.

1 May 1928

Western Canada Airways Ltd. absorbs Pacific Airways.

19 July 1928

Northern Aerial Minerals Exploration registers the first Canadian-owned and -operated Fokker Super Universal, G-CARK.

August 1928

Commercial Airways Ltd. of Edmonton begins operations.

28 August 1928

C. H. ("Punch") Dickins leaves Winnipeg with Colonel. C. D. H. MacAlpine for a 3,956-mile exploration flight across the Barren Lands.

1928

C. H. ("Punch") Dickins receives the McKee Trans-Canada Trophy for his Barren Lands flight.

January 1929

H. A. Oaks and T. M. (Pat) Reid fly the first winter flights into the Hudson Bay region.

2-3 January 1929

W.R. ("Wop") May and J. V. (Vic) Horner deliver badly needed diphtheria serum from Edmonton to Fort Vermilion, Alberta, in a wheel-equipped, open-cockpit Avro Avian.

3 January 1929

H. A. ("Doc") Oaks and T. M. (Pat) Reid in a Fairchild FC-2W-2 and a Fokker Super Universal leave Cochrane, Ontario, and reach Richmond Gulf on the east coast of Hudson Bay.

23 January 1929

C. H. ("Punch") Dickins leaves Fort McMurray, Alberta, in a WCAL Fokker Super Universal to establish an air service to Fort Simpson, Northwest Territories.

1 May 1929

A. S. Dawes obtains the registration CF-AEC for the first Canadian-owned Bellanca CH-300 Pacemaker.

14 May 1929

Western Canada Airways registers the first Junkers W. 34, CF-ABK, to be imported into Canada.

17 May 1929

C. S. (Jack) Caldwell becomes the first Canadian member of the Caterpillar Club when he is forced to parachute from an out-of-control Vickers Vedette.

1 July 1929

Flying a WCAL Super Universal, C. H. ("Punch") Dickins reaches the western Arctic coast by air for the first time.

27 July 1929

W. L. Brintnell flies a WCAL Fokker Trimotor nonstop from Vancouver to Winnipeg.

9 September 1929

Two aircraft on a mineral survey flight headed by C. D. H. MacAlpine land at Dease Point on the Arctic coast of the Northwest Territories, out of fuel, triggering the largest air search mounted to that time.

4 December 1929

Members of the lost MacAlpine party are brought back by air to civilization after they had walked to a Hudson's Bay Post at Cambridge Bay, Northwest Territories.

10 December 1929

Commercial Airways of Edmonton operates an airmail route down the Mackenzie River.

1929

W. R. ("Wop") May receives the McKee Trans-Canada Trophy.

27 June 1930

Canadian Airways Ltd. (CAL) is incorporated.

5 September 1930

Walter E. Gilbert (pilot) and Stan Knight (engineer) fly in a Canadian Airways Ltd. Super Universal over the North Magnetic Pole.

4 November 1930

E. J. A. ("Paddy") Burke's Junkers F.13, CF-AMX, missing since 11 October, is discovered on the Liard River, Northwest Territories. Two passengers survive, but Burke dies of exposure.

15 May 1931

Canadian Airways Ltd. purchases Commercial Airways of Edmonton.

15 July 1931

Mayson & Campbell Aviation Ltd. is incorporated at Prince Albert, Saskatchewan. M & C would manufacture the most widely used ski pedestals in the North.

26 October 1931

Canadian Airways Ltd. registers the famous Junkers Ju. 52/3m, CF-ARM. Test flown on 27 November, 'ARM would remain the largest aircraft in Canada until 1938.

1931

The McKee Trans-Canada Trophy is awarded to George H. R. Phillips for forest-fire protection work with the Ontario Provincial Air Service.

30 January 1932

Mackenzie Air Services Ltd. is formed at Edmonton.

11 October 1932

Bud Starratt registers the first aircraft, a de Havilland Gipsy Moth, CF-AGX, to be operated by his firm, the Northern Transportation Co. of Hudson, Ontario.

May 1933

The Manitoba Government Air Service begins operations with five ex-RCAF Canadian Vickers Vedettes.

26 May 1934

Capreol & Austin Air Service imports the first Waco Standard cabin biplane, CF-AVN, into Canada and commences operations out of Toronto.

11 July 1934

Wings Ltd. is formed to operate in northern Manitoba.

30 August 1934

Grant McConachie's United Air Transport begins operations out of Edmonton.

31 October 1934

Dale S. E. Atkinson test-flies the prototype (and only) Fairchild Super 71 at Longueuil, Quebec.

1 January 1935

C. H. ("Punch") Dickins and W. R. ("Wop") May are awarded the Order of the British Empire (OBE).

29 August 1935

Starratt Airways and Transportation Ltd. is formed from the Northern Transportation Co.

14 November 1935

W. Jack McDonough tests the prototype Noorduyn Norseman, CF-AYO, the first all-Canadian bushplane.

16 May 1936

Capreol & Austin Air Service becomes Austin Airways.

16 September 1936

A. M. (Matt) Berry locates an RCAF Fairchild 71 flown by Flt. Lt. S. W. Colman, missing in the Northwest Territories since 17 August. Coleman and his engineer were safe.

1936

The McKee Trans-Canada Trophy is awarded to A. M. (Matt) Berry for his bush flying and rescue work.

16 January 1937

Grant McConachie changes his company's name to Yukon Southern Air Transport Ltd.

23 May 1937

F. I. Young, flying a Custom Waco for Dominion Skyways, introduces a scheduled air service between Montreal and Rouyn, Quebec.

5 July 1937

Grant McConachie inaugurates a regular air service between Edmonton and Whitehorse with his Ford Tri-motor, CF-BEP.

August 1937

H. Hollick Kenyon penetrates deep into the Arctic searching for missing Russian flyer S. Levanevsky and his crew.

2 February 1938

W. J. (Jack) Sanderson test-flies the Fleet 50 Freighter for the first time at Fort Erie, Ontario.

27 October 1938

Ginger Coote Airways is incorporated at Vancouver.

14 February 1939

Canadian Airways carries out tests of the Worth oil dilution at Sioux Lookout, Ontario, using Junkers W.34, CF-AQW.

27 November–20 December 1939

Bill Catton (pilot) and Rex Terpening (engineer), using a Canadian Airways Junkers W.34, CF-ASN, carry out the longest ambulance flight on record, from Winnipeg, Manitoba, to Repulse Bay, Northwest Territories, and return.

1 August 1940

Canadian Airways begins the first flights of the Manuan Airlift. Over 3,000 tons of freight are eventually carried from Beauchene to Manuan, Quebec, for the construction of power dams.

May 1940

Nickel Belt Airways begins operations out of Sudbury, Ontario.

1940

T. W. (Tommy) Siers receives the McKee Trans-Canada Trophy for his work on the Worth oil dilution system.

14 June 1946

A. M. Mackenzie test-flies the prototype F-11 Husky.

1946

Group Captain Z. L. Leigh receives the McKee Trans-Canada Trophy for 1946 for his 20-year contribution to civil aviation.

16 August 1947

Russell Bannock test-flies the prototype DHC-2 Beaver at Downsview, Ontario.

1950

Carl C. Agar is awarded the McKee Trans-Canada trophy for his pioneering work in the development of helicopter operations in Canada.

12 December 1951

George Neal test-flies the prototype de Havilland DHC-3 Otter at Downsview, Ontario.

30 December 1952

Canadian Pratt & Whitney Aircraft tests the first R-1340 aircraft engine to be built in Canada.

1952

Tommy Fox purchases the Bristol Freighter demonstrator, which has been touring Canada, for use by his Associated Airways.

1952

Squadron Leader Keith Greenaway receives the McKee Trans-Canada Trophy for his work in northern aerial navigation.

February 1953

Rimouski Airlines becomes Quebecair Inc.

30 May 1953

Central B.C. Airlines Ltd. becomes Pacific Western Airlines Ltd. and absorbs three of Tommy Fox's Associated flying companies.

1955

G. L. McGinnis receives the McKee Trans-Canada Trophy for his work supplying DEW Line sites during construction.

6 July 1956

A. M. McKenzie test-flies the prototype Alvis Leonides-powered Super Husky.

1956

Squadron Leader R. T. (Bob) Heaslip earns the McKee Trans-Canada Trophy for his contribution to helicopter operations during Mid-Canada Radar Line construction.

30 July 1958

George Neal and David Fairbanks test-fly the de Havilland DHC-4 Caribou prototype at Downsview, Ontario.

1 August 1960

Stan Haswell test-flies the prototype Found FBA-2A at Malton, Ontario.

25 October 1960

The Hon. J. Angus MacLean opens the National Aviation Museum at Uplands Airport, Ottawa.

30 May 1961

R. H. (Bob) Fowler and George Neal test the Canadian Pratt & Whitney PT6, the first Canadian turbo-prop engine, in a modified Beech 18 at Downsview, Ontario.

1961

W. W. ("Weldy") Phipps receives the McKee Trans-Canada trophy for his work in northern flying and his development of oversized tires for tundra operations.

17 September 1962

Mrs. J. A. Richardson purchases the last operating Junkers W.34, CF-ATF, and presents it to the National Aviation Museum.

12 June 1963

Found Brothers test their FBA-2C at Malton, Ontario. This very sturdy aircraft would be used in limited numbers throughout the Canadian north.

31 December 1963

R. H. (Bob) Fowler test-flies the de Havilland DHC-2 Mk 3 Turbo Beaver.

1963

Frank A. McDougal receives the McKee Trans-Canada Trophy for developing new techniques of forest-fire suppression.

20 May 1965

R. H. (Bob) Fowler test-flies the de Havilland DHC-6 Twin Otter prototype at Downsview, Ontario.

May 1967

De Havilland Canada sells the last of a production run of 466 Otters to Laurentian Air Services of Ottawa.

27 November 1967

The RCAF's four surviving Bristol Freighters are retired and sold to operator Max Ward for use on Wardair's northern air freight operations.

August 1968

The (Found) Centennial 100 receives Type Certification, but the aircraft does not perform as well as the Found FBA-2C from which it was derived, and only three are built.

22–23 June 1970

Canadian Forces C-130 Hercules aircraft from 435 Squadron

airlift 200 troops to Wood Buffalo National Park in Alberta to fight forest fires. DHC Buffaloes of 429 Squadron provide support.

1970

Transair begins to operate four Armstrong-Whitworth Argosy freighters, each powered by four Rolls-Royce Dart turbo-prop engines.

12–28 May 1971

Three CF Voyageur helicopters of 450 Squadron lay in fuel caches between Cambridge Bay and Shepherd Bay, Northwest Territories, to support summer survey operations.

Summer 1971

The vast James Bay hydro project gets under way using available heavy float planes from across Canada. Larger wheel-equipped aircraft are introduced as soon as airstrips are built.

21 September 1971

No. 440 Detachment, CF, becomes operational at Yellowknife with two Twin Otters, as the only permanent CF force in the Arctic.

27–30 April 1974

CF Hercules aircraft conduct the first large-scale paradrop at the North Pole as part of Exercise Frozen Tusker, practice for Arctic search and rescue operations. 413 Squadron Labrador helicopters remove the troops.

1976

Austin Airways purchases the first of seven Avro 748, twin-Dart-powered airliners for its northern operations.

September 1978

A de Havilland DHC-3 Otter is successfully converted to take the P & W C PT6A-27 turbo-prop engine, shedding its reputation for being underpowered.

Glossary

Ab initio

Literally "from the beginning." Refers to the first, basic phase of instruction that a pilot trainee receives.

Aero Commander

A high-wing, twin-engine aircraft with a good performance.

Aileron

A control surface: moveable sections of the trailing edge of an aircraft's wing. Ailerons on opposite wings move in opposite directions causing the aircraft to roll around its fore-and-aft axis, facilitating turns.

AME

Licensed aero-engine mechanic.

Amphibian

An aeroplane capable of operating from land or water with wheels that retract into the hull or floats.

Approach

The manner in which an aircraft is brought in for a landing.

Approach plates

Cards containing information on approaching different airports.

Avgas

Aviation gasoline, of much higher octane rating than that used in cars.

Bird dog

A light, twin-engine aircraft used for spotting forest fires and guiding water bombers.

Blades (15 or 10)

The number of propeller blades that must be manually rotated to clear oil from the bottom cylinders of a radial aircraft engine. *See* **Pull-through.**

Blow pot

A gasoline-burning device for warming aircraft engines in winter, not unlike a plumber's blowtorch.

Bollard

A stub, usually of metal, to which a mooring line can be attached.

Bump (on ski operations)

Jolting a ski-equipped aircraft to break skis loose when they have become frozen to a surface of snow or ice. To prevent such freezing, the skis are usually run up on small logs.

Bump (when flying)

The jolt felt by those in an aircraft as it passes through a vertical air current.

Cessna 180

One of a popular line of high-winged, all-metal, single-engine monoplanes.

Cessna 337

A light, low-winged, twin-engine aircraft with good performance.

Chordwise

The width of an aircraft wing across the shorter dimension (from front to back).

Civil Operations

Flying activities such as fisheries patrol and forest fire spotting undertaken on contract by the RCAF for various provinces as well as the federal government.

Come a cropper

To slip and fall (or misjudge a landing) with disastrous results.

Cowl

The streamlined metal covering fitted around an engine to control air flow and improve cooling. Also called cowling.

Crabbing

The attitude of an aircraft when it is facing into the wind but following a course several degrees to one side. If the machine were flown directly along the course, the wind would cause it to drift.

Cutthroat

A species of trout with markings suggesting a serious wound.

Dam Buster Raid

A famed World War II bombing raid in which the RAF used purpose-designed bombs that would bounce across the surface of the water in front of a dam. When the bombs hit the dam they settled down the face to a predetermined depth where the concussion of their explosion would do the most damage.

Dead-stick

To land an aircraft without power, i.e., glide down to a forced landing with a dead engine.

Detent

A stop governing the travel of a control.

Devil's club

A tall plant with sharp spines that grows in swampy areas near the British Columbia coast.

Dope

The paint used on aircraft, especially those with fabric covering.

Dory

A small, flat-bottomed fishing boat with pointed bow and stern.

DoT

The Department of Transport. A controlling body for civil aviation in Canada (now Transport Canada).

Elevators

Control surfaces on an aircraft's horizontal tailplane. Operated by moving the control stick or wheel back or forward, they cause the aircraft to climb or descend.

Fabric

The material (linen at the time of most of the stories in this book) used as the external covering for the wood or metal structure of an aircraft. It has been largely replaced by metal in modern aircraft.

Feathering (of props)

Rotating the blades of an aircraft's propeller so that their width parallels the general flow of air and they are not affected by its passage.

Flaps

Control surfaces that can be lowered in the airstream to slow an aircraft during landing and to bring the nose down.

Floats

Boat-shaped, they support an aircraft on water. Also called pontoons.

Flt. Lt

Flight Lieutenant (RCAF equivalent of the army rank of Captain), the rank below **Squadron Leader** and above **Flying Officer**.

Flying boat

An aircraft with its lower fuselage designed as a hull, enabling it to operate from water.

Flying Officer

The rank above **Pilot Officer** and below **Flight Lieutenant**.

Formate

To fly in formation with another aircraft.

Free traders

Fur traders who are independent of the Hudson's Bay Company or the smaller competing firms.

Fuselage

The body portion of an aircraft.

Gills (cooling)

Small, louvered flaps located at the rear of aircraft engines that can be opened to draw additional cooling air through the engine.

Gimbal-mounting

A means of ensuring that sensitive instruments such as an aircraft's compass are not affected by the attitude or motion of the machine.

Gross weight

The total weight of an aircraft when it is fully loaded with fuel and cargo or passengers.

Hardtack

A thin, very hard, and dense biscuit about the size of a dinner plate used as emergency rations. Not unlike a ship's biscuit.

HBC

Hudson's Bay Company.

Herman Nelson

A commercial heater used to warm aircraft engines and interiors on winter operations in the North.

Hitting down

Falling of rain or hail from a storm cloud.

Home-built

An aircraft that has been constructed from a commercially obtained kit of plans. While such a project usually requires years of spare time, it can prove considerably less expensive than purchasing a factory-built machine.

Immelmann (turn)
An aerial manoeuvre beginning with a half-loop and completed with a roll from inverted to upright to complete a 180-degree turn.

Inertia starter
A device that, when cranked, will store the energy necessary to start an aircraft engine.

Instrument Flight Rules (IFR)
Only qualified pilots are permitted to fly IFR (out of sight of the ground), relying on information provided solely by their instruments.

Jeep heater
Presumably a small and portable gas-fired heater for warming aircraft engines. *See* **Blow pot.**

Knot
Nautical measure of distance widely used in aviation 1.15 miles.

Landplane
An aeroplane equipped with landing wheels.

Leaning (engines)
Combining less gasoline with the air to be combusted in the cylinders.

Line-squall
A wall-like and rapidly moving storm front usually marking the boundary where two weather systems meet.

LORAN
Long-range aid to navigation.

Magneto coil
A device providing the ignition spark for any internal-combustion aircraft engine.

Manifold pressure
An indication for the pilot as to the actual horsepower being generated by the aeroplane engine(s)—measured in theoretical inches (of mercury).

Mixture
The blend of fuel and air ignited in the cylinders. *See* **Leaning.**

OCA
Ontario Central Airlines.

Oil dilution

The thinning of oil for winter operation. Gasoline added to the oil allows it to lubricate a cold engine. As the engine heats up the gas evaporates and the oil returns to its normal viscosity.

Oleo leg

The part of an undercarriage containing an oil-filled, shock-absorbing cylinder.

On the step

See **Step.**

Operational Training Unit (OTU)

The phase in military air training where aircrew are introduced to simulated and real combat conditions.

Pilot Officer

The lowest commissioned rank in the RCAF.

Piper Cherokee

A light, four-place, single-engine, low-winged monoplane used for personal and sport aviation.

Piper Cub

A two-place, single-engine, light, high-winged aircraft that originated in the late 1930s and enjoyed great popularity for training and sport aviation.

Pull-through

The process of rotating an aircraft's propeller to flush out any oil that may have accumulated in the lower cylinders.

Purse-seiner

A fishing boat equipped with a net that hangs wall-like in the water and can be drawn into a circle to trap the fish within.

Quadrant (power)

That portion of the control pedestal that contains the throttle lever(s) for the engine(s).

Radial engine

An engine with the cylinders arranged in a row around the crankshaft. More powerful engines have as many as four rows of cylinders.

Registration

The government assigned groups of identifying letters (or numbers for U.S. aircraft) applied to the sides of aircraft fuselages, vertical tail surfaces, or wings. The original prefixes (followed by two variable letters) for Canadian aircraft were G-CA (civil) and G-CY (military). The next civil series used the prefix CF- with three variables while the military went to numerals. The latest series is C- and four variables.

Republic Seabee

A small high-winged, amphibious monoplane with a single-engine mounted as a pusher behind the enclosed cabin.

Ribs

The lateral members in the internal structure of a wing or horizontal tailplane.

Roll-cloud

The type of turbulent, rapidly moving cumulus cloud with strong up and down draughts that makes up a line-squall.

Rudder

A control surface, the moveable after-portion of the vertical tailplane. Used alone, the rudder will skid the nose from side to side.

Saltchuck

In British Columbia, the ocean or bodies of water connected with the shores of the province.

Scow

Flat-bottomed, square-ended boat.

Seiner

See **Purse-seiner.**

Sitka spruce

Very tall, straight-grained coniferous tree, native to British Columbia, that provides the preferred wood for aircraft wing spars.

Ski pedestal

The portion of an aircraft's undercarriage connecting it with the ski and usually containing a shock-absorbing mechanism.

Spar (wing)

The main longitudinal member in the internal structure of a wing.

Spinner

A conical or dome-shaped covering that fits over the propeller hub for streamlining purposes.

Squadron Leader

RCAF rank below Wing Commander and above **Flight Lieutenant**.

Stall turn

A spectacular manoeuvre resulting in a 180-degree change of direction. The nose of the aircraft is pulled up steeply until it is about to stall. Abrupt use of the rudder will then skid the machine around into a vertical dive.

Step

The indentation midway along the underside of a float or the hull of a flying boat. When an aircraft that is taking off attains sufficient speed to climb "on the step," it is ready to leave the water.

Stick

Joystick or control column. The principal means by which the pilot operates the control surfaces of an aircraft. Sideways movement operates the ailerons while fore and aft movement operates the elevators.

Stinson 108

A light, fabric-covered, high-winged cabin monoplane first built in the 1930s and occasionally used for short-range taxi operations in the bush.

Strut

A part of an aircraft's structure, usually exposed, that supports or separates the wing(s) or makes up the undercarriage, usually of a streamlined section.

Tanker (aerial)

A water bomber.

Tour (RCAF service)

The number of combat missions flown by a fighter pilot or bomber crewman before he stepped down for leave and a new posting.

Trim (on control surfaces)

Small moveable sections on control surfaces that may or may not be adjustable in flight and control the aircraft's attitude in the air (nose-up or nose-down).

Unstick
> Breaking contact with the ground or water as an aircraft takes off.

Vedette
> A light, open-cockpit, wooden-hulled flying boat of biplane configuration with a single pusher-mounted engine, used by the RCAF for aerial photograhy, fire spotting, and general duties.

Visual Flight Rules (VFR)
> Apply to pilots who do not have IFR (Instrument Flight Rules) qualifications and must fly within sight of the ground at all times.

V-1
> Velocity one, the point at which a pilot must decide to take off— or not.

Water rudders
> Small metal control surfaces hinged to the back of seaplane floats or to a flying boat's hull to help the pilot steer when his aircraft is afloat.

Wobble pump
> The small, one-handed pump, operated with a back and forth motion, used to transfer gas from drums to an aircraft's fuel tanks.

Suggested Reading

The books in the following list are devoted wholly or in part to bush flying. Those marked with an asterisk (*) are autobiographical accounts.

Allan, Iris. *Bush Pilot.* Toronto: Clarke-Irwin, 1966.

Avery, Norman. *Whiskey Whiskey Papa.* Ottawa: privately published, 1998.

*Balchen, Bernt. *Come North with Me.* New York: E. P. Dutton, 1958.

Beaudoin, Ted. *Walking on Air.* Vernon, British Columbia: Paramount House, 1986

*Billberg, Rudy. *In the Shadow of Eagles.* Seattle: Alaska Northwest Books, 1992.

Blanchet, Guy. *Search in the North.* Toronto: MacMillan, 1960.

Bruder, Gerry. *Heroes of the Horizon.* Seattle: Alaska Northwest Books, 1991.

Bungey, Lloyd M. *Pioneering Aviation in the West.* Surrey, British Columbia: Hancock House, 1992.

Burton, E.C., and R.S. Grant. *Wheels, Skis and Floats.* Surrey, British Columbia: Hancock House, 1998.

Callison, Pat. *Pack Dogs to Helicopters.* Vancouver: privately published, 1983.

*Cilbert, Walter E., and Kathleen Shakleton. *Arctic Pilot.* Toronto: Thomas Nelson, 1940.

Condit, John. *Wings over the West.* Madeira Park, British Columbia: Harbour Publishing, 1984.

Constable, George, ed. *The Bush Pilots.* Alexandria, Virginia: Time Life Books, 1983.

Corley-Smith, Peter. *Bush Flying to Blind Flying.* Victoria: Sono Nis Press, 1993.

———. *Barnstorming to Bush Flying.* Victoria: Sono Nis Press, 1989.

Corley-Smith, Peter, and David Parker. *Helicopters: The British Columbia Story.* 2nd edition. Victoria: Sono Nis Press, 1998.

*de Goutiere, Justin. *The Pathless Way.* Vancouver: Graydonald Graphics, 1969.

Ellis, Frank H. *Canada's Flying Heritage*. Toronto: University of Toronto Press, 1954.

———. *In Canadian Skies*. Toronto: Ryerson Press, 1959.

Finch, John, et al. *Wings Beyond Roads End*. Regina: Saskatchewan Education, 1992.

Foster, J. A. *The Bush Pilots*. Toronto: McClelland & Stewart, 1990.

Fuller, G. A., J.A. Griffin, and K.M. Molson. *125 Years of Canadian Aeronautics: A Chronology 1840–1965*. Toronto: Canadian Aviation Historical Society, 1983.

Godsell, Philip H. *Pilots of the Purple Twilight*. Toronto: Ryerson Press, 1955.

*Grant, Robert S. *Bush Flying*, Surrey, British Columbia: Hancock House, 1995.

Hotson, Fred. *de Havilland in Canada*. Toronto: Canav, 1999.

———. *The de Havilland Canada Story*. Toronto: Canav, 1983.

*Jones, Marsh. *The Little Airline that Could, Eastern Provincial Airways*. St John's, Newfoundland: Creative Publishers, 1998.

Keith, Ronald A. *Bush Pilot with a Brief-Case*. Toronto: Doubleday, 1972.

*Leigh, Z. Lewis. *And I Shall Fly*. Toronto: Canav, 1985.

*Leising, William A. *Arctic Wings*. New York: Doubleday, 1959.

Lunny, June. *Spirit of the Yukon*. Prince George, British Columbia: Caitlin Press, 1992.

Matheson, Shirlee S. *Flying the Frontiers*, Calgary: Fifth House, 1994.

———. *Flying the Frontiers*. Vols. II and III. Calgary: Detselig, 1997, 1990.

*McLaren, Duncan. *Bush to Boardroom*. Winnipeg: Watson and Dwyer, 1992.

Milberry, Larry. *Air Transport in Canada*. Vols I and II. Toronto: Canav, 1997.

———. *Austin Airways*. Toronto: Canav, 1985.

———. *Aviation in Canada*. Toronto: McGraw-Hill Ryerson, 1979.

Mitchell, Kent A. *Fairchild Aircraft 1926–1987*. Santa Ana, California: Narkiewicz & Thompson, 1997.

Molson, K. M., and A. J. Short. *The Curtiss HS Flying Boats*. Ottawa: National Aviation Museum, 1995.

Molson, K. M., and H. A. Taylor. *Canadian Aircraft Since 1909.* Stittsville, Ontario: Canada's Wings, 1982.

Molson, Kenneth M. *Canada's National Aviation Museum.* Ottawa: National Aviation Museum, 1988.

————. *Pioneering in Canadian Air Transport.* Winnipeg: J. A. Richardson and Son, 1974.

Myles, Eugenie L. *Airborne from Edmonton.* Toronto: Ryerson Press, 1959.

Oswald, Mary, ed. *They Led the Way.* Wetaskiwin, Alberta: Canada's Aviation Hall of Fame, 1999.

*Parsons, H. P. "Hank." *Trail of the Wild Goose.* Winnipeg: privately published, 197(?).

Pickler, Ron, and Larry Milbery. *Canadair: The First 50 Years.* Toronto: Canav, 1995.

Potter, Jean. *The Flying North.* New York: Doubleday, 1946.

Reid, Sheila. *Wings of a Hero.* St. Catharines, Ontario: Vanwell, 1997.

Render, Shirley S. *Double Cross.* Toronto: Doubleday, 1999

————. *No Place for a Lady.* Winnipeg: Portage & Main, 1992.

Rossiter, Sean. *The Immortal Beaver.* Vancouver: Douglas and McIntyre, 1996.

Sandwell, Capt. A. H. *Planes Over Canada.* Toronto: Thomas Nelson, 1938.

Satterfield, Archie. *Alaska Bush Pilots in the Float Country.* New York: Bonanza Books, 1979.

Shaw, Margaret Mason. *Bush Pilots.* Toronto: Clarke-Irwin, 1962.

Sullivan, Kenneth H., and Larry Milberry. *Power: The Pratt & Whitney Canada Story.* Toronto: Canav, 1989.

Sutherland, Alice Gibson. *Canada's Aviation Pioneers.* Toronto: McGraw-Hill Ryerson, 1978.

*Theriault, George. *Trespassing in God's Country.* Chapleau, Ontario: Treeline, 1994.

*Turner, Dick. *Wings of the North.* Surrey, British Columbia: Hancock House, 1980.

Vachon, Georgette. *Goggles, Helmets & Airmail Stamps.* Toronto: Clarke-Irwin, 1974.

West, Bruce. *The Firebirds.* Toronto: Ontario Ministry of Natural Resources, 1974.

Wheeler, William J. *Images of Flight.* Toronto: Hounslow, 1992.

*White, H., and H. Spilsbury. *The Accidental Airline.* Madeira Park, British Columbia: Harbour Publishing, 1988.

*Woollett, W. "Babe." *Have a Banana.* Vancouver: Turner-Warwick, 1989.

Index

In this index, numbers appearing in italic bold type indicate photographs.